CROWOOD METALWORKING GUIDES

BRAZING AND SOLDERING

CROWOOD METALWORKING GUIDES

BRAZING AND SOLDERING

RICHARD LOFTING

THE CROWOOD PRESS

First published in 2014 by
The Crowood Press Ltd
Ramsbury, Marlborough
Wiltshire SN8 2HR

www.crowood.com

British Library Cataloguing-in-Publication Data
A catalogue record for this book is available from the British Library.

ISBN 978 1 84797 836 3

Disclaimer
Safety is of the utmost importance in every aspect of the workshop. The practical procedures and the tools and equipment used in engineering workshops are potentially dangerous. Tools should be used in strict accordance with the manufacturer's recommended procedures and current health and safety regulations. The author and publisher cannot accept responsibility for any accident or injury caused by following the advice given in this book.

Acknowledgements
I would like to thank my family for their help in writing this book, in particular my son William Lofting and my niece Bethany Old, who posed for some of the photographs, my wife's aunt, Audrey Peters, for reading through the text, and of course Pam, my wife, for the endless cups of tea. Thank you.

All photographs by Richard Lofting.

Typeset by Servis Filmsetting Ltd, Stockport, Cheshire
Printed and bound in Malaysia by Times Offset (M) Sdn Bhd

Contents

Introduction

Brazing and soldering are essentially the same joining technique, the difference between the two being the temperature at which each method is performed. In essence, a joint is made in metal using an alloy of two or more metals to, in effect, hot-glue the parts together. The word 'glue' here does not refer to just sticking something together with a sticky substance, as the items to be joined will be bonded at a molecular level to the alloy, which imparts considerable strength to the joint. The weakness will be in the strength of the brazing or soldering alloy, although the strength of some alloys that are used for filler rods are very near to the strength of the materials being joined.

In general terms, the hotter the melting temperature of the alloy, the stronger the joint. The weakest joints are made with soft solder, where the soldering alloys melt at temperatures below 450°C; in fact, some alloys will melt at extremely low temperatures and even in hot water, but these are beyond the remit of this book as they are not used for soldering. Historically, most soft solders were predominantly made from lead. However, the lead content has lately been substituted for other alloying metals, such as tin due to health concerns regarding the use of lead and its poisonous effect on the human body and of course the environment.

The middle ground is held by hard soldering, or silver soldering, where traditionally the soldering alloy contains a significant proportion of silver; details of the actual alloy contents will be given in a later chapter. Hard/silver soldering covers a temperature range of approximately 450°C through to around 850°C. As can be seen, this is already significantly higher than soft soldering, with joints effectively being made with a blow-lamp as these temperatures are beyond a soldering iron's capability.

At the top of the temperature range is brazing, with a heat range of approximately 800°C through to 1,000°C and sometimes higher. In this process, the alloying components in the filler rod are mainly copper, tin and zinc; copper and tin form the alloy bronze, while copper and zinc form the alloy of brass, both very common alloys in everyday use. While these temperatures may seem hot, they are still a fair way below welding temperatures, which are required to melt and then fuse the component parts together; typically for mild steel this is around 1,300°C.

In reality, there is no division between silver soldering and brazing – you cannot say that on one side of a temperature line it is silver soldering and the other side of the line it is brazing. All the preparation is the same, from thorough cleaning through to heating; the only difference is in the alloys being used to complete the joint. The technique of silver soldering and brazing covers an enormous field in modern engineering, as it enables the joining of dissimilar metals. For example, it is possible to braze hard tungsten carbide tips into a steel circular saw disc or lathe tool, thus giving the best of both worlds – the tips are hard and so retain their cutting edge for longer, but the centre disc or tool holder is made from steel, giving the flexibility to stand the forces produced while cutting. With the relatively new important materials like ceramics entering the field of engineering, new ways to join these have had to be devised. Originally, they were given a metallic coating to which brazing alloys would adhere, but by accident it was discovered that with an active element in the alloy, usually titanium, used in the right atmosphere, ceramic substances could be brazed directly.

With the increasing temperature range that has just been shown, so the relative strength of the joint correspondingly increases with it. This is due, mainly, to the alloying metals used in the brazing rods or sticks of solder, as stronger materials generally have a higher melting temperature than the weaker, or softer, ones. This is the derivation of the terms 'hard' and 'soft' solder. The soft solders contain lead and tin, while the hard solders contain harder metals in the alloys such as silver. Once past the hard soldering metals and into brazing, the metals used in the alloys contain brass, which in itself contains copper and zinc. At the top of the temperature and strength range alloys using nickel are used. An in-depth overview of soldering and brazing alloys will be looked at in Chapter 1, along with the fluxes that are available and their attributes.

A soft-soldered lap joint. Soft soldering is performed at the lowest of the temperature range, but will give a moderately strong and air-tight joint.

Brazing is the strongest joining technique, without melting the parent metal; this allows dissimilar metals to be joined.

Particularly with hard soldering and brazing, where a brazing torch/blowlamp is to be used, this must be carried out in a safe environment, away from inflammables. The hotter the temperature used for the process, the more oxides are produced from not only the parent metal but also the alloy bearing rods. Chapter 3 is dedicated to building a brazing hearth. Although this sounds rather grandiose, it is nothing more than a frame made from angle iron or sheet steel with the working area lined with commercially available firebricks.

When heating metals, or anything else for that matter, the increase in temperature has the effect of increasing the production of surface oxidation, so in all forms of soldering and brazing a flux is required. In general terms, a flux keeps the metal from oxidizing during heating and removes any that has already formed, from both the item being joined and the soldering/brazing rod. The exception is specialist brazing rods that are available for joining copper to copper, which are in effect self-fluxing due to their phosphorous content.

The technique of lead loading automotive body panels with lead solder has now been, mainly, superseded by the use of plastic fillers, although there is a niche in the vintage and classic car world where originality is important. Plastic fillers are very good, being easy to mix and apply and then contour to the desired shape, but they are not as permanent as lead loading. A higher skill level is required with lead loading, as the vagaries of the alloy are exploited during the application process, spreading the lead alloy like butter with wooden paddles, unlike polyester fillers, which are mixed with a hardener; the only skill here is to get the mix on to the panel before it sets. Once again, the poisoning effect of the lead content rears its ugly head. The finish cannot be done with power tools, as this will fill the air with lead particles, but is done by hand with body files and abrasives.

At the end of the book, Chapter 9 is dedicated to electrical soldering. Although it is basically just soft soldering, there are several criteria that need to be considered to produce a satisfactory electrical joint.

Alloy rods and fluxes required for soldering and brazing operations.

1 Alloys and Fluxes

The three categories of soldering and brazing are:

◆ soft soldering, which is mainly carried out with a soldering iron and low temperature alloys
◆ silver soldering, also known as hard soldering, which is performed in a brazing hearth using alloys containing silver
◆ brazing, which is also performed in a brazing hearth, but using higher temperatures and higher temperature melting alloys, producing the strongest joints.

SOLDER AND BRAZING RODS

To help understand the process of soldering it will be advantageous to look at the composition of the alloys used in the rods for each process. First, the basic question needs to be answered: what is solder? Traditionally, soft solder was mainly based on alloys predominantly made of lead, but with the increasing awareness of the poisonous effects of lead on persons and the environment, lead solders are no longer used in plumbing. In addition, lead-based electrical solders are now virtually banned worldwide due to the possibility of the lead content leaching into water supplies from landfill rubbish sites, as a result of the ever increasing amount of discarded electronic equipment. European Union directives are now in place to prohibit the use of lead in various industrial processes and the construction industry, with it being replaced by other non- or less poisonous metals such as tin, copper and silver as the main alloy. Apart from the attributes that lead gave to solder alloys, such as good wettability, it was a cheap and readily available commodity, whereas the alternatives are relatively more expensive to source and extract from the ore containing the metal.

Another question is why use alloys for the filler material instead of a pure metal? There are several reasons. A pure metal melts at a higher temperature than an alloy containing the same metal and by combining two or more metals the attributes from each can be taken advantage of in the alloy. Also, varying the mixture will vary the melting temperature and other factors such as wettability or the tensile strength of the resultant solder alloy.

As already seen in the introduction, soldering has been neatly divided up into three sections, namely soft soldering, silver (or hard) soldering and brazing. These are arbitrary boundaries that have been historically developed over the years. As far as the alloys used for all three categories are concerned, there are no delineating boundaries, as the only difference between any of the alloys used for soldering or brazing is the temperature at which they melt and consequently can be used at, and their service life. It will be explained later on in this chapter that most alloys have two temperatures given in their specification, unless

Soft solder is available as sticks or wire on a reel; most soft solders are predominantly non-lead based today, due to environmental concerns.

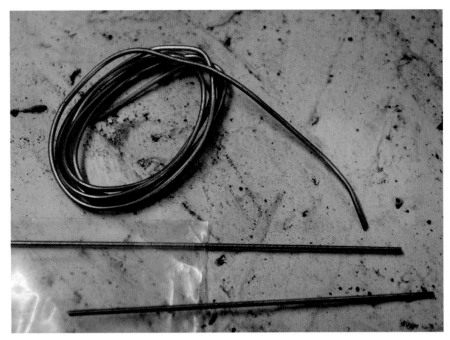

Silver solder comes as wire or rods depending on the diameter being used and for what purpose; thin foils are also available for specific purposes.

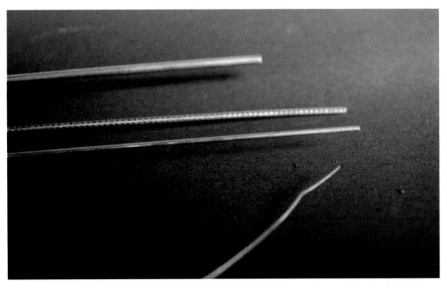

Brazing rods are very much like silver soldering rods, the difference being the melting temperature of the alloy they contain and that they are usually brassy coloured.

they are a eutectic alloy and therefore have a single melting point.

NAME THAT ALLOY

There are various ways that manufacturers of soldering and brazing rods describe their particular soldering and brazing alloys, sometimes using the main alloying metal's percentage with a range description, for example Silver-flo 40. This alloy, which is from the Johnson Matthey range, has a silver content of 40 per cent, copper 30 per cent, zinc 28 per cent with 2 per cent tin, a melting

> **THE DEFINITION OF AN ALLOY**
>
> The term 'alloy' describes a mixture of two or more pure metals. The most common feature of an alloy is that its melting temperature is always lower than any of its constituent parts. Solders use the differing alloying contents mainly to determine the melting characteristics of the solder. Alloys are not confined to soldering and brazing, but have many uses, such as brass made from copper and zinc, bronze made from copper and tin, and of course the world would be a different place without the steel alloys that are commonly used. Alloys have many uses in producing properties in materials, such as hardness and wear-resistance, which the parent materials do not possess.

range of 650–710°C and a tensile/shear strength of 450/155 N/mm^2. This particular alloy mix meets the widely recognized international standards and is variously known as AG20, L-Ag 40 Sn and AG105, depending upon whether you are looking at the BS1845, DIN 8513 or the EN 1044 standards. Or the alloy's name may contain the temperature as part of its description, for example Indalloy 281, this being a eutectic alloy having a single melting temperature of 281°F or 138°C. The use of the Fahrenheit temperature scale in the description usually means that the manufacturer is from America, where the Fahrenheit scale is standard. Where a manufacturer produces a range of alloys in a series it makes it relatively easy to pick various alloys from the range so that soldering in a sequence or step soldering can be performed in confidence, without disturbing the previously made joints. This very useful technique will be covered in detail in a later chapter.

ALLOY MELTING TEMPERATURE

A feature of an alloy is that it does not usually go directly from a solid to a liquid, as does a pure element such as copper, although there are exceptions to this. Two terms are used in defining the melting and solidifying of an alloy – these are the liquidus and solidus points.

Liquidus

The liquidus point in an alloy describes where the whole alloy is deemed to be a homogenous liquid without any remaining solid crystals from any of the constituencies of the alloy. This is the upper temperature quoted for a particular alloy. For example, a 5 per cent silver (Ag), 95 per cent tin (Sn) soft solder is quoted in data charts as having a melting temperature range of 221–235°C; in this case, the liquidus point is 235°C. As the liquid alloy cools down, the liquidus point marks where crystals once again start to form within the liquid alloy, with the amount of crystals increasing as the temperature falls. Once cooled to the solidus point, the alloy is once again a solid.

Solidus

The solidus point is the start of a range at which an alloy will melt, as different crystals within the alloy will melt at differing temperatures. As in the example above, the lower figure of 221°C given in the range for that particular soldering alloy is the solidus point. The alloy will not be completely melted until the liquidus point is reached; this is when the whole alloy is a homogenous liquid.

The difference between the liquidus and solidus points can be small or relatively large, depending on the alloying materials. Within the range of the two points, the alloy will become a mush, not unlike melting snow, and is neither completely solid nor a liquid. Normally during soldering and brazing operations it is nothing more than an inconvenience, but movement of items being soldered during this phase will possibly result in a dry or porous joint. However, as will be seen later in Chapter 8, this mushiness can

be used to our advantage. For example, when lead loading an automotive body panel with body solder, if the alloy is kept between the solidus and liquidus points, it can be shaped and contoured, not unlike plastic. But of course should a little bit more heat be applied, so that the liquidus point is reached or superseded, the body solder will be all over the floor!

Liquation

As stated above, generally the solidus and liquidus range is nothing more than an inconvenience while soldering. However, with some alloys that have a large temperature range and particular alloying elements a problem called liquation can occur. This is where on reaching the solidus temperature, as the alloy begins to melt, the prolonged heating cycle can cause the alloying elements to separate. The ones with the lower melting temperature melt first, leaving crystals of the remaining alloys behind. Once this has occurred, the only effective remedy is to allow everything to cool and then clean off the separated constituents of the alloy, and the inevitable surface oxides, and start all over again. In order to prevent liquation, the best solution is to complete the joint as quickly as possible by rapid heating of the items being soldered or brazed through the solidus and liquidus points. Apart from the unsightly appearance of a joint where liquation has occurred, the resultant joint will possibly be brittle and porous and thus unreliable in service.

Eutectic Alloys

The exception to an alloy with both a solidus and liquidus point is a eutectic alloy. This type of alloy has a single melting point, with the solidus and liquidus points being effectively one and the same, as in a pure element such as iron or copper.

Although not connected to soldering, it is of interest to note that low-temperature

eutectic alloys are used in the fusible plug within automatic fire-prevention sprinklers, whereby an alloy with a melting temperature of below 100°C melts in the heat of an emerging fire and so releases water from the sprinkler. One such alloy, named Wood's Alloy after the American physicist Robert W. Wood, melts at 70°C. It contains 50 per cent bismuth (Bi), 26.7 per cent lead (Pb), 13.3 per cent tin (Sn) and 10 per cent cadmium (Cd), although it is detrimental to health as it contains lead and cadmium. Another safer eutectic alloy is Field's Alloy, named after scientist Simon Quellen Field, which has a melting temperature of just 62°C and is an alloy of 32.5 per cent bismuth (Bi), 51 per cent indium (In) and 16.5 per cent tin (Sn).

ALLOY SELECTION

All sorts of base metals are used in solder and brazing alloys, with the main ones being shown in the table. Some relatively rare or exotic metals are used in small amounts; these give the alloy certain advantageous qualities, such as wettability of the solder, which means that the solder will flow on to the item being soldered with ease.

The Internet will reveal a huge range of silver soldering and brazing alloys, with some interesting proprietary names, such as Silvaloy 5, Matti-sil, Sil-Fos 56, Argo-braze 40N and so on. Often a clue to the main alloy in the rod is in the name, for example Silvaloy; here, the main alloying component is silver. However, this is not always the case; therefore, before undertaking any serious silver soldering or brazing it is a good idea to have a look at the manufacturers' data sheets, which are freely available online and elsewhere. This will allow an informed decision to be made on the suitability of the rods to be used for a specific job where exacting standards are required.

Most manufacturers have a range, or several ranges, of rods with specific alloy content but, by varying the amounts, small changes in the alloy mix can have profound

Alloying Metals

Name	Chemical symbol	Melting temp. °C	Benefits when added to solder or brazing alloys
Antimony	Sb	631	increases alloy strength
Bismuth	Bi	271	lowers melting point in alloy, improves wettability
Cadmium	Cd	322	good wettability, restricted use due to health effects
Copper	Cu	1,085	lowers alloy melting point, improves wettability
Gold	Au	1,064	wets most metals, but expensive
Indium	In	157	lowers melting point, improves ductility
Lead	Pb	327	solder base metal, but alternatives used due to environmental concerns
Manganese	Mn	1,246	improves wettability on carbides containing titanium
Nickel	N	1,455	enhances high-temperature chemical and physical properties
Phosphorous	P	44	free flowing, good capillary penetration
Silicone	Si	1,414	gives high-temperature strength and oxidation resistance
Silver	Ag	962	gives alloy mechanical strength, improves resistance to fatigue
Tin	Sn	232	alternative to lead in solders, good wettability
Zinc	Zn	693	lowers melting point, low cost

common case would be the need for correct rod selection when joining stainless steel fittings where these are going to be in contact with, in particular, salt water. If the right rods are not used, what is known as crevice or interfacial corrosion will occur between the stainless steel and alloy interface, causing possible porosity or weakness due to one of the alloying metals within the alloy being attacked by the salt.

When deciding what alloy composition of the filler rod to choose for the task in hand, be it soft soldering, silver soldering or brazing, it must be noted that all alloys weaken before signs of melting occur. This has implications in the serviceability of a joint, especially if the joint is subject to heat under working conditions. Most alloy data sheets will give the service temperature limits of individual alloys.

Most silver solders are now produced to the ISO 17672 standard, with fresh virgin materials being used rather than reclaimed materials so as to meet this standard. It is particularly important when building a pressure vessel, such as a steam boiler by silver soldering, for example, that solder

effects on the alloy's abilities, such as wettability, flow characteristics and gap filling. The other main effect is that of changing the solidus and liquidus temperatures, which may be critical for the metal being joined, or to the service to which the soldered or brazed joint is to be put.

When selecting a suitable alloy filler rod, there is more to it than just finding one that melts in the correct temperature range. Other factors come in to play, such as the suitability of the alloying elements within the rod. For example, it is no good using a solder containing lead when soldering is undertaken connecting the gold wires on an integrated circuit chip to the external connections on the case, because the lead component will leach away the gold from the delicate wires. While this may be a somewhat rarefied example, a more

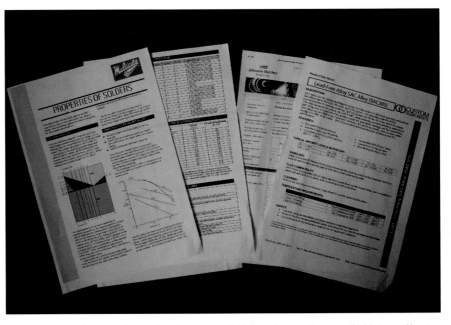

For anything other than general work, data sheets are available from the manufacturers of soldering and brazing alloys. These detail the constituents and characteristics of the alloys and, importantly, their melting temperatures.

made to this standard is used along with the correct grade of copper. This will avoid any possibility of intermetallic compounds forming that will cause embrittlement of the silver soldered joint and the surrounding area, by alloying with any metals such as aluminium and titanium contained within the silver solder. Failure of the boiler joints and seams in this example would have huge safety implications to not only you, but to any bystanders in the vicinity, rather than be just a disappointment. The usual course of events during boiler construction is to employ a hydraulic test; this entails filling the boiler completely with cold water, then pressurizing it to above the working pressure at which the boiler is designed to work. This will indicate any leaks within the boiler, but, more importantly, will not cause a catastrophe, should one of the seams fail. Water, being virtually incompressible, will lose pressure immediately upon a seam failure. If air or steam were to be used, being compressible a lot more expanding energy would lie behind the failing seam and therefore be a lot more dangerous.

Different sized jobs require different rod sizes.

Size Matters

One factor that needs consideration, especially with silver soldering and brazing work, is selection of the correct size of filler rod diameter. As already stated, the rods come in all shapes and sizes and indeed some have specific specialist uses. As most of the filler is drawn into the joint, from this perspective the rod size will have little effect on the finished joint, which will take so much filling alloy and that will be that. As a general rule, when working with thin materials such as sheet work, the smaller sized rods or even wire are ideal; if a thicker rod were to be used it would take more heat from the items being heated to melt the alloy, causing a pause in the alloy flow and resulting in a patchy joint. If larger sections are being joined, which in themselves will have more stored heat in them, this is the

time for using thicker rods. If thinner rods or wire were used on the larger job, while the joint would be successfully completed, it would take a lot of rods or wire and, with the thinner section rods being more expensive pro rata, the cost of the job will escalate. In an ideal world, we could stock up on the common grades used and in all diameters, but this would also be somewhat expensive.

If buying rods for a specific job and there are a few rods left over, label them so that at a later date you will know what the alloy content of the rods is, as one rod looks much the same as the next. The only discernible feature between various rods might be a slight difference in colour; for general work this may not matter, but for use in the construction of pressure vessels it may be critical to a safe job.

Solder and Brazing Rod Storage

The above statement leads nicely into rod storage. The best storage for solder and brazing rods is a dry, warm place away from the dirt in the workshop; plain rods will not be affected by moisture as the flux-coated

ones will be, but any moisture on the rods will allow the general detritus, grinding dust and so on floating around the workshop to stick to the rods. This will not hurt the rods themselves, but could lead to contamination of the joints made using them. The flux coating on the flux-coated rods will deteriorate if allowed to remain in a damp atmosphere for any length of time. If buying in bulk, the rods will usually come in a protective plastic tube and if the lid is kept on this will keep most contaminants at bay during storage. If purchasing smaller amounts, storage in a plastic container would be ideal.

HEALTH CONCERNS

As has already been noted, lead is poisonous to the human body and the environment, a fact that is now well known. Cadmium has traditionally been used in silver solders as it imparts very good flow characteristics to a silver solder alloy, making it easy to use. The downside is that as an alloy containing even small quantities of cadmium will give off dangerous fumes as it is heated. There is a vast range of alloys now available that

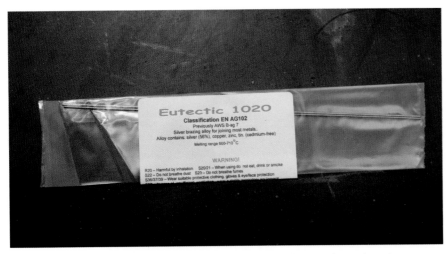

At a minimum, store your rods in a bag to keep off contaminants and label them so that you know the composition of the rods for future use. Once the data is lost and the rods muddled up, it will just be guesswork.

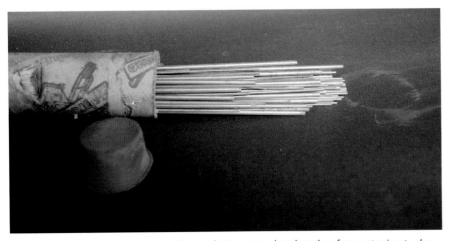

An ideal way to store your brazing/silver soldering rods is in a stout tube to keep them from contaminants, also, keep the lid on when not in use.

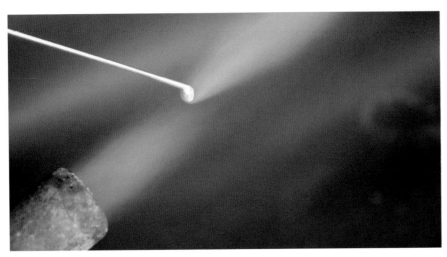

Do not use the heat from the flame to melt the rod, as this will inevitably lead to failed joints.

do not contain cadmium and it would be sensible to use an alloy from these ranges so as to avoid the risks from cadmium fumes. If it is found that an alloy has to be used that contains cadmium, then all the correct fume extraction should be in place before attempting to use these rods. If using rods of an unknown source and the exact alloy content cannot be verified, it would be prudent to assume that they contain cadmium and take the correct precautions.

SILVER SOLDERING

One of the problems associated with silver soldering and indeed brazing, is applying

Use the heat in the items being joined to melt the rod, as this will ensure penetration by capillary action, as the alloy is drawn into the joint rather than being melted on top.

A technique that can be used, especially on decorative work, is to place small pieces of solder along the line of the joint before heating. This will ensure that a minimum amount of solder is used, avoiding getting it all over the place and of course saving money by using less solder.

the filler rod before the parts to be joined are hot enough. The joint may look to be at the right temperature, that is, glowing red hot, but it may have not quite reached the melting temperature of the alloy being used. As the rod is advanced into the joint it melts in the heat of the flame from the blow-lamp, not from the heat of the parts being joined. Providing more heat is applied until the silver solder is seen to be drawn into the joint by capillary action, there will be no problem. However, if the alloy is left sitting on the top surface, this will achieve nothing other than an ugly mess.

A useful technique, especially when doing work of a decorative nature such as jewellery, is to cut small pieces of silver solder and place them at points along the joint to be made, with the application of flux only where the solder is intended to run. As the whole assembly comes up to tempera-ture, the pieces of solder will be seen to flow into the joint very neatly, requiring very little in the way of cleaning. This technique will be demonstrated in Chapter 5.

In industrial applications, silver solder is available in the form of preformed rings, foil strips and so on, so the right amount of solder is already at the joint before heating. This not only keeps things neat and tidy, it also makes the use of the silver solder more economical. When making many joints, it is surprising just how much solder will be used, with the price of the silver content in the alloy always attracting a high price.

STEP BRAZING

Step brazing will be covered in detail later in the book, but an overview of the differ-ent alloys will be given here. As an example, a model steam boiler would be made from many components needing to be joined together. To do this all at once would be all but impossible, as some parts would require brazing before others and with prolonged heating many brazing alloys will lose their attributes. When a steam boiler is made, the pressure vessel (i.e. the boiler itself) would require brazing first and then ancil-lary components will then be required to be attached later. If the same brazing alloy was

to be used for all the joints, on heating the original brazed joints would remelt, making construction very problematic.

The process of step brazing is fairly simple, in that all that is required is careful selection of the brazing rods to be used, with the alloy that has the highest melting temperature being used first, followed by the second highest and so on, until the whole assem-bly has been completed. Alloys with fairly short melting temperature ranges are ideal for this process, as this allows alloys to be chosen that do not overlap. Of course, this is a simplified overview, as in reality there are all sorts of issues to contend with for a satisfactory result. For example, there would also need to be careful selection of fluxes, as these also have an optimum temperature range in which they work and prolonged heating can have a detrimental effect on their efficiency.

BRAZING OR SILVER SOLDERING OF CARBIDE TIPS

This is a most useful process with the advent of tungsten carbide being formed into teeth for industrial and indeed domestic power tools, such as lathe and milling machine cutters. These teeth or cutting bits can be sharpened to a very fine edge and, being extremely hard, last a long while between sharpening and can be used without cutting fluids, as the carbide retains its hardness at elevated temperatures, unlike the tool steel traditionally used to make these cutting tools. With the advent of holders made for these cutting tips, it has removed the neces-sity of silver soldering or brazing them on to a tool holder, although the process is still used for things like circular saw blades. Foils of alloy are available with a thin foil strip of copper or other softer metal within, which is then placed between the tip and the tool body. The strip of softer metal acts as a buffer between the hardened tip and the steel tool body, absorbing the shock loading that could crack the hardened carbide if it

To take advantage of the hard characteristics of tungsten carbide, which will keep a sharp edge for longer, tips are brazed on to a steel disc to make a circular saw blade.

A carbide-tipped lathe tool has many advantages over a high-speed steel one, for example, heat produced during turning does not affect the hardness of the tip. Higher speeds can be employed, increasing output, and harder materials can be machined without problem.

was silver soldered or brazed directly to the tool body.

FLUXES

The word flux originates from the Latin word *fluxus*, meaning to flow, and this is what a flux assists with – it allows the filler alloy to flow on the metal surface as well as removing oxides. Today, most fluxes are made and categorized under the ISO 9454-1 standard; this not only gives the minimum you should expect from a flux, but also consistent results from a particular flux. As already discussed, the flux required for any soldering or brazing operation is essential to a successful outcome in the production of a joint between two items.

Before any attempt is made to solder or braze, the parts to be joined require a thorough cleaning to rid them of any contaminants that may stop the joint from being made, such as paint, rust, oil or grease. Under this layer of surface coatings, there will possibly be an oxide layer, the thickness of which will depend upon several contributing factors, including how well the item has previously been cleaned, what metal it is and what environment it has been kept in. However thick this oxide layer is, it will need to be removed for a successful soldering or brazing operation. Mechanical removal of the surface oxides alone will not allow successful soldering to take place, as upon heating to bring the parts up to soldering temperature fresh oxides will be produced on the surface. This is where the use of a flux comes into play.

When any metal object is subjected to being heated, the surface oxidation of that object increases as the heat rises. It therefore follows that for a successful joint to be made, the oxidation needs to be eliminated, or at least stopped and removed, so that the solder or braze can flow easily into the capillary joint. The flux needs to be active in the right temperature range at which work is being carried out in order to be effective.

For most general purpose soft soldering work, a mild flux paste is all that is necessary to clean the work and keep the oxides at bay during heating and subsequent soldering.

Once into the realms of silver soldering and brazing, with the higher temperatures employed, flux usually comes as a powder; this can be used 'as is', or mixed to a paste and applied before any heating begins.

If working on used items, for example in repair work, sometimes a more aggressive flux, such as Baker's Soldering Fluid, will be required to move stubborn oxides that build up over time on the metal surface.

If materials are heated without flux, the surface will soon show signs of oxidation (black scale), making any sort of sound joint impossible.

When heating with the flux applied, it effectively seals the surface from the oxygen present in the air, prohibiting it from combining with the material and stopping oxides from forming.

Soft soldering fluxes are available in a range of cleaning ability, the very mild being nothing more than an inert material keeping oxidation at bay, through to the very aggressive fluxes that contain acid in varying degrees. These will clean off surface contaminants as the soldering progresses, as well as protecting the joint from the air; however, this must not be seen as a substitute for thorough mechanical cleaning before the soldering operation, as there is a limit to what these fluxes will remove. The downside of using a flux containing acid is that once the flux has done its job it will still keep eating away at whatever it has been placed on until the acid element is used up or neutralized. Without going into the chemistry behind the facts of acids and alkalis, the acid needs to be washed off or, at least neutralized, with some form of mild alkali substance to prevent further corrosion.

For electrical soldering, where any acid in the flux would corrode, a naturally occurring

Fluxes suitable for medium-temperature silver soldering and high-temperature brazing therefore need to be made from differing materials so as to remain integral and active at the different temperatures. Soft soldering fluxes are predominantly derived from organic materials, but once the higher temperatures of silver soldering and brazing are reached, the organic nature of these materials will inevitably mean that they will not withstand the high temperatures and will burn or disintegrate. Higher-temperature fluxes are therefore predominantly mineral-based.

Soft Soldering Flux

For soft soldering work, fluxes usually come in liquid or paste form, as the working temperature range is fairly low. This will allow the flux to stop any oxidation from when it is applied to the joint, right up to the melting temperature of the solder and just beyond.

Typically, just before placing plumbing joints together, a small amount of flux is applied; twisting as they go together is all that is required to spread the flux around the whole joint area.

As the joint reaches the correct temperature, the flux will start to bubble at the joint interface. Applying solder now, it will be drawn into the joint by capillary action, making a watertight pipe joint and leaving a silver line all round the joint line.

After cleaning, the joint will give trouble-free service for many years to come.

Once the joint has cooled down, it can be seen that there is a flux residue left on the joint; this needs to be cleaned off with water, as the flux will remain acidic.

flux, rosin, is used; this is derived from various species of pine tree. At normal temperatures rosin has the appearance of a glassy solid and in this state it is virtually inert, being non-corrosive and non-reactive, but at soldering temperature the mix of organic acids within clean copper effectively and keep it oxide-free while the solder bonds to it.

Silver Soldering Flux

The fluxes available for silver soldering usually come in powder or paste form and are generally made from fluorides or fluoroborates, which are alkali. These have to perform their function at higher temperatures than their soft soldering cousins and therefore are made from different material. The powders are either mixed with water to form a paste (the addition of a few drops of washing-up liquid will help with wetting the surface as the flux is applied), and then applied to the joint just before heating, or the brazing rod is heated and dipped into the powder so that a blob of flux melts and adheres to the rod. In some quarters, this technique is known as 'hot rodding'. It is then applied to the joint as heating takes place from the end of the rod; further flux on the rod melts as the filler rod melts. As the flux melts, it will be seen to have a wet appearance where it has been applied. It stops the oxidation from occurring by covering the area against the effects of the atmosphere. Any oxides already formed will be lifted by the active ingredients within the flux and will float to the top of the flux, leaving a chemically clean surface into which the solder or braze will bond.

Brazing Fluxes

At the higher temperatures used for brazing, again different materials are required, although, as stated above, there is little to distinguish between the top end of silver

General-purpose brazing flux, in this case Sifbronze. The downside of this flux is that it sets hard like glass and can be difficult to remove without resorting to mechanical means.

soldering and the lower end of brazing and the same is true of the flux used. Borax has been the traditional flux for brazing operations and is still a good one for ordinary steels where brass-based alloy fillers are to be used. The main downside of using borax or borax-based fluxes is that once they cool down they set to a glass-like state that can sometimes be difficult to remove. The effective option is to break it up mechanically, that is, hitting it with a pointed instrument and then briskly wire-brushing to remove the loosened particles.

Flux-Coated Rods

Silver soldering and brazing rods are available with a flux coating. This has the advantage for general work of saving time, in both the preparation and execution of the job. As the items to be joined are heated up, the flux-coated rod is touched lightly to the joint area. This will melt a small amount of flux from the rod, which will protect against oxidation as the heating continues to brazing temperature. Once it is deemed that the temperature has been reached, the rod

Silver solder flux from the Easy-flo range; more aggressive types are available for use on stainless steel and so on.

Flux-coated rods are available in small quantities from some of the larger DIY stores, just right for the odd brazing job.

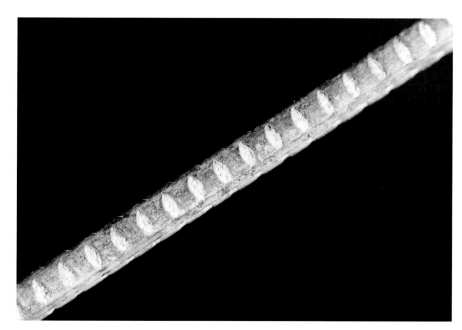

For general workshop joining, brazing rods are available with the flux attached to the rod, in this case pockets rolled into the rod itself. This provides enough flux from the rod during use to keep the joint oxide-free. As the parts are heated the rod is touched to the joint, whereby the flux will melt, protecting the joint until the alloy temperature is reached. As the rod is fed in, more flux will be added.

red hot, this will harden the steel, making it impossible to carry out any further cutting or filing of the components. Cooling by quenching in water has the effect of cracking the brittle flux and most of it will drop off, leaving the remains to be removed with a wire brush or other mechanical means. It is important to remove any residue, as once the item is painted or otherwise finished, any remaining flux will have the tendency to detach itself some time in the future, taking the protective paint with it and exposing the joint to possible corrosion.

Self-Fluxing Rods

So far in this chapter it has been stated that to perform an integral soldered or brazed joint, flux is a requirement. The exception is alloy rods, which are available specifically for joining copper. These contain phosphorous in the alloy, which reacts with the oxygen

is touched to the joint line again. This will add more flux and if the temperature is right the alloy will begin to flow into the joint. It will pay to keep the flux-coated rod out of the direct path of the torch flame, as this will tend to melt the flux from the rod, giving too much flux in one place and not enough in another. It has to be said that flux-coated rods are only suitable for general work; for anything with a deep capillary joint, it will be found that it is better to flux the components before assembly and use a plain rod, adding more flux where necessary as the job proceeds.

The flux that is applied is only effective at removing oxidation for a limited period and may require replenishing as the work proceeds. Once the work is complete the flux will solidify, with the higher-temperature types setting into a glass-like substance, hard and brittle. As the soldered/brazed item cools down, if it is made from mild steel or copper, it can be quenched in water once the solder/braze has solidified. If anything more than mild steel is quenched while it is

High-temperature brazing will leave a hard flux residue on the brazed joint, which will probably require mechanical effort to remove.

that ordinarily produces the oxides and prevents them from forming. These rods can also be used on other metals if used with a flux, although not on iron and nickel, as the phosphorous can cause intermetallic compounds to form at the joint boundaries, leading to brittle joints.

Cleaning Off the Flux

Once the soldering operation has been successfully completed, the residue and spent flux will require removal to facilitate a decorative finish and to stop any further action by acid-bearing fluxes. Some water-soluble fluxes after use will remain hydroscopic, that is, they will absorb moisture from the air even though the finished joint is dry after completion. This will possibly reactivate the acid component within the residue flux and start corrosion.

Most soft soldering fluxes will only require washing off with some warm water and a good scrub with a stiff brush, with perhaps a wash-over with a mildly alkaline solution to neutralize any remaining acid. Bicarbonate of soda, available from most supermarkets and cooking stores, will be ideal for this, but under no circumstances 'borrow' the pot from the kitchen: workshop contamination will be detrimental to your health if it gets into your food.

Medium-temperature silver soldering fluxes, such as the Easy-flo range from Johnson Matthey, will be removable after soaking in water for 30 minutes or so and then scrubbed with a stiff brush, before rinsing with more clean water.

The higher-temperature brazing fluxes will set like glass once cooled and, as stated above, can be partially removed by quenching as the item cools. This is only suitable if the materials making up both the item and the alloy are not sensitive to quenching. Another way to remove this hard flux residue is to chip it with a pointed tool, not unlike chipping away welding slag after arc welding, and then wire-brushing to remove the loosened particles. Any stubborn remaining residue may require grit blasting to remove the last traces where it is critical to be completely clean.

Flux Safety

Safety once again rears its head with fluxes, mostly when they are in powdered form. Whey they are dry and at normal temperatures, fluxes are all but inert. However, when the temperature is raised during the soldering process the constituencies of the flux can vaporize and it is these fumes that cause concern. On the Internet, and in particular on some of the forums, there are discussions about the safety aspect of the fluxes available and the dire consequences of their use. Even the natural rosins associated with electrical soldering have given rise to concerns, with asthma-type symptoms being reported by some during use. Obviously liquid and paste fluxes with an acid content are a potential hazard if in contact with the skin and vaporized acid will not do the lungs much good either. Provided that safety observations already undertaken during the heating and use of the alloys are adhered to and adequate ventilation is provided, and in addition inhalation of the rising plume of smoke/vapour is avoided, you should come to little harm from the fluxes used.

Storing Flux

As was seen with the storage of flux-coated brazing rods, a dry, warm atmosphere is the best place for the storage of fluxes. The powdered types, although they usually come in a plastic container, need to be kept dry until use. If moisture is absorbed into the powder through leaving the lid off, over time the flux will set into a solid mass. For the same reason do not mix the flux with water in the flux container, but use a separate vessel for the mixing. When hot rodding, that is, dipping the hot brazing rod into the flux, be careful that on dipping the rod into the flux container you do not push too hard and cause the hot tip of the rod to penetrate through the bottom or side of the flux container. This will not only lead to the flux trickling out of the hole, therefore wasting the flux, but can also lead to moisture from the air entering the container, again setting the contents into a solid lump.

Lower-temperature soldering fluxes will wash off with water; some may require soaking for 30 minutes to soften.

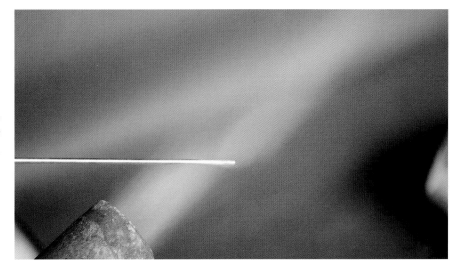

A simple process that can be used with plain rods is 'hot rodding', which entails heating the end of the rod in the torch flame.

Once deemed hot enough (not too hot or the rod will melt), push the rod tip into the flux powder for a second or two.

On withdrawal from the flux, a blob of flux has adhered to the rod tip. As this is used up as you work, just re-dip the rod tip into the flux when more flux is required.

For any soldering or brazing operation a form of heat will be required.

2 *Soldering Irons and Blowlamps*

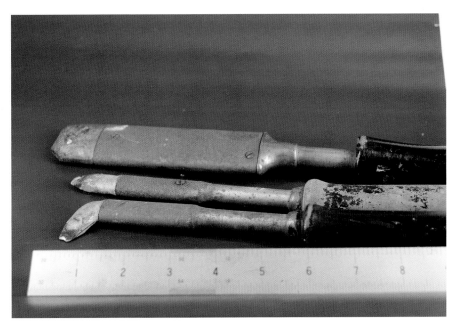

For the larger soft soldering job, large electric soldering irons are available, up to 200–300 watts.

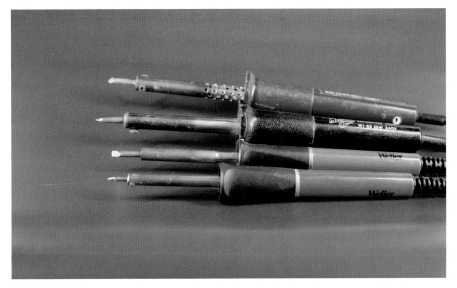

Small electric soldering irons are the correct tool for electrical soldering; the miniature ones, typically 9 watts, are ideal for circuit board work.

At the very heart of soldering, or brazing for that matter, a heat source is required to bring things up to the melting temperature of the filler rod being used. In this book, we are concentrating on the options available in a general or home workshop, although in industry many different ways are employed to heat components for soldering and brazing. An example is the induction coil, whereby an electric current is passed through a coil, which in turn induces heat from within into the items to

Gas-powered blowlamps come in all shapes and sizes, from this small general-purpose type through to larger gas torches, which are powered by industrial-sized cylinders.

be joined, making it a quick and efficient method, but one that is only cost-effective when used for mass-production. The basic requirement of the heat source is to bring the component parts and the joint area up to the melting temperature of the solder or brazing rod fairly quickly so as to avoid oxide formation. Soldering in some forms and brazing are carried out with just a flame, but for normal soft soldering a soldering iron will usually be needed. This will give an efficient transfer of heat to the item being soldered, with the minimum of oxide formation.

THE TRADITIONAL SOLDERING IRON

Historically, a soldering iron, with the working bit usually made of copper riveted to an iron shaft with a wooden handle, was heated in a coke or coal fire, with the soldering iron placed inside a piece of iron pipe or similar. This avoided the corrosive effects of the fire attacking the soldering iron bit and consequently the joint to be made, and also lessened the formation of oxides on the soldering iron away from the direct heat of the fire. Once the bit of the soldering iron was up to the required temperature to melt the solder, it would be mechanically cleaned and then dipped into the flux before applying solder to 'tin' the tip of the soldering iron. During the soldering process, as the heat from the soldering iron was given up into whatever item was being soldered, the soldering iron would require reheating in the fire before carrying on with the job. This may have taken several reheats until completion of the work. If more than one soldering iron was available, one iron would be heating while the other was being used. Once paraffin was freely available and the use of pressurized blowlamps was in vogue, soldering became easier and more portable, with repair work being undertaken on site rather than having to be taken to the coke furnace, or a furnace having to be set up

Before propane and butane torches, the paraffin blowlamp was the main heating source for soldering. This entailed heating the torch itself to vaporize the paraffin. Once hot enough and alight, the paraffin was pressurized by the built-in pump, creating a hot flame. Enough heat was available for small silver soldering jobs, but not quite enough for brazing.

Copper soldering irons were available in many shapes and sizes for use with the paraffin blowlamp. Of course, these can still be used with its modern gas equivalent.

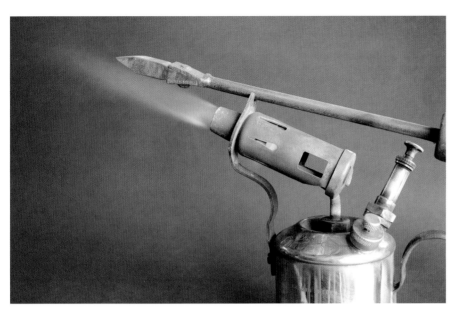

Using copper soldering irons is straightforward. The soldering iron is heated by the flame away from the tip. When hot enough, the flux will bubble, indicating it is time to add the solder to tin the iron.

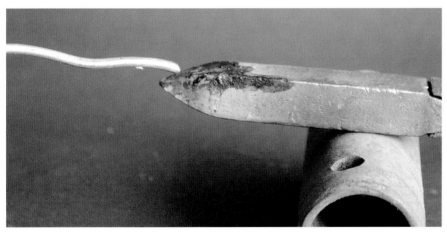

To use the iron successfully, the tip needs to be tinned adequately with solder.

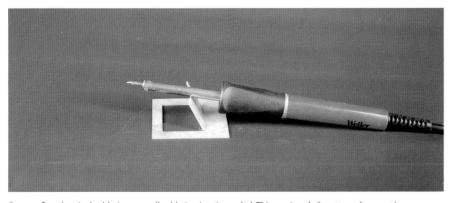

For very fine electrical soldering a small soldering iron is needed. This one is only 9 watts and can get in some very tight spaces.

on site. This was especially welcomed with lead work being common on roofs during this period, and water and drainage pipes also being made from lead, which lends itself to easy soldering, making watertight joints.

SOLDERING IRONS

When undertaking soldering work with a soldering iron, although the temperature of the soldering iron is important, it is imperative to select the right size of bit. Too small and the heat will be lost too quickly. This will mean that as the heat is drawn into the object being soldered, either not enough heat will be available to complete the joint in one go, or the soldering iron will require frequent reheating to do the job properly. A large soldering iron bit will retain more heat for longer and is ideal for the larger job, but will require more time to heat up initially and at subsequent reheating, so will consume more fuel to carry out the same job.

ELECTRICAL SOLDERING IRONS

The above paragraph is just as relevant when applied to the electrical soldering iron. The only difference is that the heating coil within the soldering iron will not be able to keep up with the heat output that is required for the job in hand and consequently the solder will solidify on the tip of the iron. The only remedy here is to use a bigger wattage soldering iron. Electrical soldering irons are rated by the wattage that they consume, from the small ones with a wattage of 8 or 9 watts for fine electrical/electronic soldering, up to large ones of 120 watts or so capable of soldering sheet metal. Soldering irons are now available with instant heat being available as soon as they are switched on and upon release of the switch go off again. These are fine for where intermittent work is being carried out, saving power.

For heavier work, larger electric soldering irons are available. The front two are 60-watt irons, the one at the back is 240 watts.

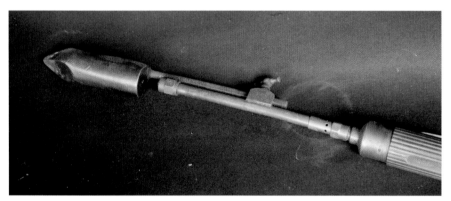

To avoid having to keep reheating the iron, propane-heated soldering irons are available. Once the soldering temperature has been reached, continuous soldering can be undertaken without having to keep reheating.

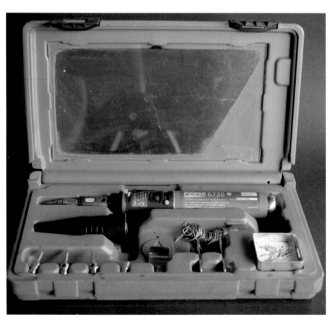

A neat kit containing everything for the small soldering job – a butane-powered blowlamp, with various soldering bits, electrical solder and a tip cleaning sponge, all in a carrying case.

GAS-POWERED SOLDERING IRONS

Where more heat is required for the larger job, or the operation is being carried out far from an electrical power source, a gas-heated soldering iron should cover your needs. These work by the back of the soldering bit being heated by a gas flame. Some gas torches are adaptable to this mode of operation by the addition of a purpose-made adapter and a specially designed bit to fit it. This heats the iron throughout the entire operation. The heat input is easily varied to suit the job by adjusting the flame and different-sized bits are readily available.

BUTANE-POWERED SOLDERING IRON

One innovative product to come on the market over the last few years is the butane-powered blowtorch/soldering iron. The design incorporates a refillable gas cartridge in the handle of the device with a small blowtorch at the front end. The fancy ones incorporate a piezo crystal lighting device so that the whole thing is self-contained and does not require matches to light it. Various attachments are usually supplied in a blow-moulded plastic case, one of which is a soldering iron bit. Some just heat the back of the iron's bit with the flame of the blowtorch, but the more sophisticated models have an inbuilt catalyser, which produces heat from the gas without a flame. This heats the back of the soldering iron bit instead of a bare flame, giving greater efficiency and therefore longer use between gas refills. The process when using a catalyst is to light the blowtorch as usual with the ignition device until the catalyst is sufficiently warm. The flame is then extinguished, usually by a mobile shutter that momentarily cuts across the gas flow, and the gas burns without a flame as it interacts with the catalyst.

Filling of the gas canister is simple. All that is required is a butane lighter refill canister available from most large supermarkets or tobacconists, then selecting the right adapter to go between the refill canister and the blowtorch filling orifice. With the refill canister held with the filler down, press the two together, which will allow the liquid butane in the canister to flow into the blowtorch cartridge, until the pressure of both equalizes. For obvious reasons, all gas filling operations need to be done in the open air and of course well away from any sources of ignition, such as naked flames or pilot lights.

CARE OF THE TIP

As has been observed in this book, when heating any metal it has the effect of increasing the natural oxidation on the metal's surface. The bit on the soldering iron is no exception, hence the importance of properly tinning the bit in the first place and subsequently making sure that the tinning is in good order as you proceed during soldering. Apart from the tip oxidizing, interactions between the solder used and the soldering iron's tip can readily occur. This usually takes the form of depletion of material from the iron's tip, causing it to distort and erode away. The manufacturers of soldering irons have tried to alleviate this by substituting the traditional copper tip with ones made from iron, but the heat transfer is not quite as good as with copper, so the wattage of the iron needs to be slightly higher than with a copper bit.

Occasionally, it is a good idea to look at the soldering iron tip; any signs of distortion or erosion will need to be remedied in order to keep the soldering iron in tip-top condition. During a soldering operation, the usual procedure is to wipe off the iron tip every so often to remove any dross on the tip, that is, oxides which have accumulated from the joint being made, the solder itself and of course the soldering iron tip. If this dross is left to accumulate to any degree it will start to impede the operation, by insulating the heat from the iron and slowing its

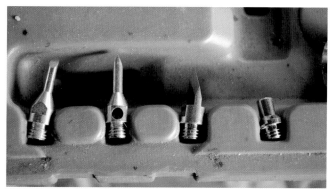

With these soldering and air heating bits all aspects of light soldering are catered for.

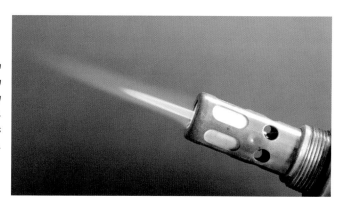

This type of butane torch is ideal for small heating jobs without the soldering iron attachment, just using it as a gas blowlamp.

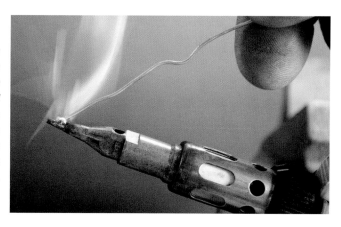

The same torch with the soldering iron attachment; here the tip is heated with the aid of the built-in catalyst, converting the gas to heat without a flame. As there are no wires, it is fully portable.

It is important that before use a soldering iron tip is cleaned, so that it will tin easily. On this large soldering iron, a file is used to get back to clean copper.

The easiest way to flux the soldering iron tip is to dip it straight into the pot of flux momentarily; this will ensure that any lingering oxides are removed.

After being in use for a while the soldering iron tip will become eroded.

Before the flux boils off, reheat the tip and melt solder on to it. An even coating where the flux has been means that the tinning has been successful.

The best way to bring the tip back to shape is to use a small file until the tip is flat once more, removing any eroded pits.

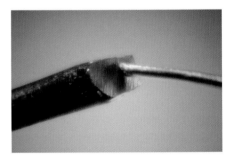

Before the tip has time to oxidize, heat immediately after cleaning, tin and apply solder, in this case electrical multi-core solder as it is a small iron for electrical work.

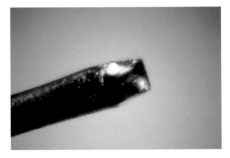

The tip of the soldering iron retinned ready for more soldering.

transfer into the work. We are trying to keep the joint free from oxidation, so to use a tool encrusted in it defeats the object.

Retinning the Tip

The retinning of the soldering iron tip is a simple operation. All that is required is to flux the tip before reheating and as soon as the iron is up to temperature, apply some solder to the tip. Wiping the tip while hot will reveal if it has been tinned completely, as the tinned area should have a silvery shine all over. If it is at all patchy, showing that the tinning has not taken properly, it will need redoing, giving extra attention to the cleaning. Obviously this same technique will be required on a soldering iron for electrical work, the only difference being that the fluxing and tinning of the soldering iron

tip is done in one operation, as the solder for electrical work comes with the flux inside the solder as a core. Do not be tempted to use an acid-bearing flux designed for other purposes on your soldering iron for electrical work, as these fluxes are much too corrosive to use on delicate electronic components and cannot be washed off without damaging the components. Any small amounts transferred from a contaminated soldering iron bit could have disastrous, or at the least indifferent, results.

Dressing the Tip

To dress the soldering iron tip, all that is required is a fairly fine file that can be drawn across the working end of the tip to take away any distortion and irregularities and bring it back to the original shape, so that it can then be retinned ready for use again. This must be done when the soldering iron is cold or has cooled down sufficiently to handle; any attempt to do this while the iron is hot will inevitably result in burnt fingers.

GAS-POWERED BLOWLAMPS

With the advent of gas-powered blowlamps, large soldering jobs became easier and quicker, as the rigmarole of lighting the paraffin blowlamp, with its lighting and warming-up procedures, could be bypassed. The gas-powered blowlamp has almost instantaneous heat and most have interchangeable nozzles, giving different heat outputs for different-sized jobs.

If you are fortunate enough to have access to a supply of piped gas in the workshop, you could install a brazing hearth with a blowtorch running from the supplied natural gas, which is methane. The usual

For most general work a medium-sized blowlamp nozzle will suffice, giving plenty of heat without using large amounts of gas.

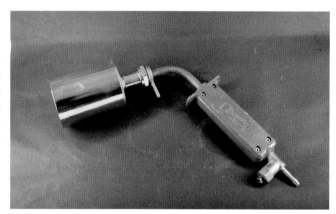

This Bullfinch nozzle No. 1270 will give copious amounts of heat, but will use gas at a rate of about 2.5kg per hour. This will raise the temperature in large objects quickly, avoiding long heating cycles.

Small blowlamps are available for the smaller job. This one uses a Campingaz cylinder, but many other makes are available, along with various gas mixtures.

method is to burn this gas with air. Commercial brazing hearths are available with a built-in compressor, which supplies air under pressure. This is mixed with the gas and, with the increased pressure, the resultant gas/air mixture burns more fiercely, giving off more heat than would otherwise be available if the gas were to be burnt at the pressure supplied by the network.

Liquefied Petroleum Gas (LPG)

Both propane and butane are LPGs, being a by-product of the distillation of crude oil, and at normal temperatures and pressures they are both gases. A convenient quality is that if each gas is compressed at normal ambient temperatures, it turns into a liquid. For propane, this is somewhere in the region of 125lb^2 in; for butane, this happens at a

---

LPG GAS PRODUCTION

A by-product of the oil industry, Liquefied Petroleum Gas (LPG) is produced when crude oil is distilled into differing factions. Apart from the obvious petrol and diesel oil, there are many different substances that are distilled from the basic crude oil that is extracted from the ground. The process is not unlike the distilling of whisky, whereby the fermented liquid containing the alcohol is heated and the alcohol within is vaporized, as it has a lower boiling point than the water in the liquid. This is then condensed back into a liquid by cooling. The water is left behind, giving us whisky, which is then collected in another vessel. The distilling of crude oil is similar, in that the oil is heated and at various temperatures the differing factions are drawn off, though of course on a much larger scale.

lower pressure. This property allows a lot more gas to be contained in a small space, such as a gas cylinder. As the gas is withdrawn from the cylinder via a pressure regulator the liquefied gas turns back to a gas, to burn at the torch as it is mixed with air or oxygen. On turning back from a liquid to a gas, both propane and butane take up on average 250 times the volume that they did as a liquid, hence a relatively small container stores a lot of gas.

Propane

Propane is a three-carbon atom alkyne gas at normal temperatures and pressures. Its chemical formula is C_3H_8, it has a boiling point of −42°C at an air pressure of 1bar and has a specific gravity of 1.52. When burnt with air, a flame temperature of 1,982°C can be reached. When burnt with sufficient oxygen, a temperature of 2,882°C can be obtained.

Propane bottles are available in a variety of sizes, from the small 3.9kg up to the very large 47kg. They all have the same standard left-hand threaded outlet.

Butane

Butane is a four-carbon atom alkyne gas. Its chemical formula is C_4H_{10}, it has a boiling point of −1°C at an air pressure of 1bar and has a slightly higher specific gravity than propane of 2. Butane can exist in two forms, ordinary butane and isobutene. The difference is in how the atoms bond with one and other within the gas. When burnt with air, butane can reach a flame temperature of 1,996°C.

From the specific gravity shown of both gases, 1.52 and 2, it can be seen that they are both heavier than air, which has a specific gravity of 1 at 1bar. This has certain safety implications that will be discussed further on in this chapter.

Propane and butane are both suitable gases for use with portable blowlamps. Butane has a slightly higher heat output, but for practical purposes this will make

Butane, liquefying at a lower pressure than propane, is available in small throwaway tins as well as refillable cylinders, but it does not readily vaporize at low air temperatures.

very little difference, although it will start to freeze in the cylinder at a higher temperature than propane. Due to the effect of expansion of the gas as it is released from the cylinder, not unlike the release of air from a bicycle tyre when frost can be seen to form on the tyre valve as the air is released, frost can similarly be seen to form on the outside of the gas cylinder when a lot of gas is being withdrawn. This has the effect of lowering the temperature of the liquid gas still in the cylinder, reducing the amount that is vaporized. This, for those who are interested, is explained by Boyle's law, below.

The fixed-pressure regulator is only good for low-temperature heating.

Boyle's Law

Boyle's law was formulated in 1662 by Robert Boyle, an eminent scientist, with the simple formula $P_1V_1/T_1=P_2V_2/T_2$, with P=Pressure, V=Volume and T=Temperature. The logic of the formula is that if the volume, in this case the gas cylinder, is constant, then the only variables can be the pressure and temperature. It follows therefore, that if you lower the pressure within the cylinder, the temperature will go down correspondingly. This explains why, when the pressure is released from a bicycle tyre, it will lower the temperature of the air causing frost to form. The same is true with our gas cylinder: if the ambient air temperature is low enough, and so consequently the liquefied gas within the cylinder is near to its vaporization point, any further reduction in temperature caused by the release of gas through the regulator will mean that further vaporization of the gas will slow and then cease as the liquefied gas goes below its vaporization point.

The same effect will be observed with propane if enough gas is withdrawn in a short space of time, although its boiling temperature is much lower at −42°C. This will also result in the regulator freezing, exacerbating the situation. The industrial solution is to use cylinder heaters to keep the regulator warm, or two or more

The adjustable regulator. This can give varying pressures depending upon the heat requirements of the job in hand. A gas pressure between 0 and 2bar can be set.

cylinders manifolded together. This has the effect of reducing the amount of gas being withdrawn from each cylinder and thus reducing the risk of freezing.

Regulators

If purchased as a kit, a gas torch will come with a suitable regulator for its use. However, there is a wide range of regulators on the market for use with commercial-sized propane cylinders. Some gas torches work at a fixed gas pressure, so regulators are available at fixed output pressures of usually 1bar (15psi), or for serious brazing where the higher pressure is essential, at 4bar (60psi). The most versatile ones are the regulators with a pressure adjuster; here, the regulated pressure can be fine-tuned to the job you are working on. Before changing the regulator, check that the torch you have

The beauty of an adjustable regulator is that it allows the operator to fine-tune the torch flame, saving gas and possible overheating of the job in hand.

is compatible with the gas pressure that is intended to be used.

Throwaway Canisters

With the lower liquefying pressure of butane, it is the ideal gas for filling the lightweight throwaway canisters of gas that are used on the smaller portable blowlamps. It is also used in many things, including lighters, ladies' hair curlers and, as we saw earlier in this chapter, mini-blowlamp/soldering irons. For these, the gas is sold in a canister with a selection of nozzles that will enable the filling of all these appliances. Sometimes a mixture of other gases is added to the butane in the throwaway cans. This has the effect of regulating the gas as it is withdrawn from the canister with the differing vaporization of the two gases.

To refill the portable gas soldering iron, a butane lighter refill can will do the job at little expense.

Piezo Ignition

A lot of gas torches available today have an ignition device built into the body of the torch to facilitate lighting. These devices contain a naturally occurring crystal, called a piezo crystal. This works by the natural phenomenon that when a piezo crystal is compressed it gives out a pulse of electricity, which is then fed to a spark gap just in front of the gas nozzle. A spark jumps across this gap and, provided that the gas is turned on, it will ignite, obviating the need to carry matches or a lighter.

What's Available

One of the benefits of using modern gas torches is that they usually have a screwon nozzle and sometimes are sold as a set, complete with a regulator and a selection of nozzles, possibly in a case to keep it all together. This, of course, means that a nozzle of a differing size can be fitted for large or small jobs without investing in another torch, plus a saving on the amount of gas being used is always worthwhile.

There are two main types of gas blowlamp. For small jobs and where portability is essential and the volume of work is low, the blowlamp with its own replaceable canister of gas would suffice. These canisters are thrown away once all the gas is used up; they are a one-use cartridge incapable of refilling, a new one being purchased and screwed on to the blowlamp body before use.

If a lot of silver soldering or brazing is to be undertaken, a blowlamp that runs from a larger cylinder of gas will be required. These are refillable on an exchange basis. Having made your initial gas purchase, you will have also paid a deposit on the gas cylinder, but when returned for a refill you will subsequently only pay for the gas. If you are workshop-oriented, with little outside work, the larger sizes of cylinder will give the best value, as the bigger the cylinder purchased,

the cheaper the gas pro rata. The largest size supplied by Calor, for example, contains 47kg of gas. The drawback with this size of cylinder will be transportation, as the cylinders are of necessity heavy and hold a lot of gas, adding more weight. Smaller sizes of cylinders are available, giving smaller quantities of gas, but of course the weight

A complete general-purpose propane blowlamp kit with various sizes of nozzle.

The smaller gas blowlamp; some are also available with different sizes of nozzle.

is less. This is particularly useful if mobility is essential, such as for a plumber moving from job to job.

SUITABILITY

For soft soldering and silver soldering the general-purpose blowlamp will suffice. These produce enough heat in the 600–700°C range, although the actual flame temperature of these torches will be significantly higher than this. For any brazing work, that is, in the 800–1,000°C range, one of the specialist propane torches that are capable of using the gas at a pressure of up to 4bar will be required. Although the flame temperature is not much above that of the general-purpose torch, it is the amount of heat that can be transferred to the job that is important. As you heat the object at one point, the heat is radiating away from the rest of it, just as fast, so a quick heat transfer from the flame is all important, especially at the higher temperatures and this is achieved partly by the higher gas pressure.

An alternative without going to the whole set-up of oxyacetylene gear is the torch that will burn acetylene with air, although of course this entails renting a cylinder of acetylene. One example is the AutoTorch produced by Bullfinch, which will produce a flame temperature of 1,100°C, but with a broader flame than the oxyacetylene welding torch.

FLASHBACK ARRESTERS

The flashback arrester is a safety device that is fitted between the gas regulator and the hose to the torch. It prevents a flashback, hence its name, travelling back from the torch up the hose and reaching the gas cylinder, where it could initiate self-combustion within an acetylene cylinder. The device works by cooling the burning gas to below the ignition temperature of the gas and a pressure-sensitive valve shuts off the gas supply if it senses that a higher pressure

exists in the hose than that coming from the regulator.

Standard propane and butane torches burning with air do not require the use of a flashback arrestor at the gas cylinder, but oxypropane, oxyacetylene and acetylene torches do.

MULTIPLE TORCHES

If heat from one torch will not produce enough, then another torch can be used at the same time to increase the total heat output. Obviously, you cannot hold two torches at once and feed in filler rods, as you will not have enough hands, but with the use of an assistant or fixing one torch in a

clamp it is possible. This will allow you to use the fixed torch for general heating with the second hand-held, allowing it to be freely moved around and giving extra heat to any areas that require it. Safety is obviously a concern with the use of a single gas torch, but when more than one torch is used, extra care will need to be taken regarding where each torch is pointing, with there being an additional gas pipe in the vicinity of the brazing hearth.

OXYPROPANE

When burning gas with air, it is the oxygen in the air that allows the gas to burn. In scientific terms, the gas burning is just a chemical

For serious heating jobs, an oxypropane torch with a big nozzle will supply more than enough heat to braze very large components. In this case, a pepper-pot nozzle is fitted to a standard oxyacetylene cutting head.

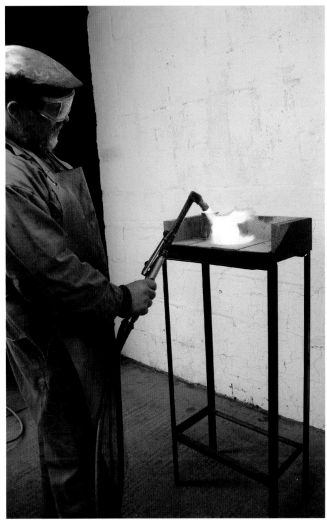

reaction between the gas and the oxygen. Air has only about 20 per cent of oxygen along with other gases such as nitrogen and carbon dioxide. From this, it follows that if the oxygen content is increased, the chemical reaction will be more intense and the temperature of the burning gas will be raised. Oxyacetylene equipment will produce a working temperature of 3,500°C, which is far in excess of temperatures required for soldering and brazing. With the correct nozzle, propane can be burnt with oxygen, albeit at a lower temperature than oxyacetylene; this will be somewhere in the region of 2,820°C, but more than enough for the task in hand. The main benefit is that large nozzles can be employed to produce greater volumes of heat that are more suitable for brazing, with temperatures around 1,000°C being required.

When propane is burnt with oxygen, differing from oxyacetylene where for a neutral flame the ratio is roughly 1:1 ratio, it will require 3.63 times more oxygen by mass to achieve a completely clean burn with only CO_2 being produced as a by-product of the combustion. Lowering the quantity of oxygen will produce soot from the excess carbon left and possibly carbon monoxide.

Once a propane mixer and heating neck has been fitted to the standard welding torch, propane heating nozzles can be fitted that will deliver copious amounts of heat from the range of nozzles available. The smallest, 1H, will at a minimum deliver 72,000 British thermal units (BTUs), with the largest, 5H, at a maximum delivering 618,500 BTUs; however, the latter will consume rather large amounts of gas. If the larger sizes are used for more than a few minutes two or more cylinders would need to be manifolded together to keep up the flow.

OXYACETYLENE

Oxyacetylene welding and cutting equipment has a multitude of uses in and around the workshop, but its use is not advised unless training in its safe use has been undertaken and the safety implications are fully understood. Details shown in this book are illustrative only and are not intended as a beginner's guide. A good starting point would be Richard Lofting, *Crowood Metalworking Guides – Welding*.

Oxyacetylene equipment could be used just as well for silver soldering and brazing. The temperature of the flame burns at a high enough temperature, but perhaps in this case it would be a drawback because the flame is somewhat concentrated. This is ideal for welding, as it requires a tightly concentrated flame, but a more diffuse, broader flame is required for silver soldering and brazing operations. With the concentrated heat from an acetylene nozzle, there is a real danger of overheating in one part, while another is underheated. The upshot of this would be that in parts the silver soldering or brazing alloy could boil off, while the underheated part would not have enough heat to draw the alloy into the joint. The resultant joint possibly would be porous and incomplete and therefore not fit for purpose. Pepper-pot nozzles, which have multiple holes, are available for use with

A set of half-sized bottles running oxyacetylene gear, the ultimate in portable heating equipment. Larger sizes of cylinder are available, but are much heavier to move around and are more expensive, too.

acetylene torches. These would be better than a welding nozzle, but unless they are your only heat source they are not recommended for silver soldering and brazing.

The smallest of the heating nozzles that are available, AHT-25, will produce a heat output of 52,000 BTUs, but can only be used intermittently from a single acetylene cylinder, as this is near to the limit of the amount of acetylene that can be withdrawn from a cylinder before the acetone containing the acetylene is withdrawn along with the acetylene. This shows itself by a yellow or green tinge to the flame and must be avoided. For the larger sizes of heating nozzle, two or more cylinders will need to be manifolded together to increase the gas available to the nozzle.

If the only option is to use oxyacetylene with a welding nozzle, try to keep the central cone of the flame away from the work, as this is the hottest part of the flame, and while heating the items up to temperature do not keep the flame in one place, but keep the flame moving all over the item. As everything heats up you will see the whole thing take on a red glow rather than at just one point, as when welding. It is recommended to use a slightly oxidizing flame when brazing, as this will help to keep the zinc element within the brazing rods from boiling off.

GAS CYLINDER SAFETY

Propane and Butane

As already stated in this chapter, gas, be it propane, butane or a mixture of both, will be stored under pressure. This will be relatively low, somewhere in the region of 125lb per square inch for propane and somewhat lower for butane. This is the pressure that the gas turns to a liquid and therefore means a lot of gas can be stored in a cylinder and the pressure will remain constant until the cylinder is almost empty, when the pressure will drop rapidly. As in common with all fuel

gases, the fittings on industrial-sized cylinders will have a left-hand thread, so that only the correct pressure regulator can be connected. This is denoted by the fittings having a notch on each corner of the nut to avoid any confusion.

Acetylene

Acetylene is an alkyne gas, formula C_2H_2. Listed as an explosive, this gas can become unstable if not treated with care. Acetylene cylinders differ from others by having a porous filling in them soaked in acetone to facilitate storage of the acetylene under any pressure. Sudden shocks, such as dropping the cylinder or banging it with an object, could cause the acetylene to break down within the cylinder. A flashback from the welding torch could cause a shockwave travelling back up the gas hose to initiate this same process, causing the cylinder to become hot as the gas breaks down. Once this process has started, the only thing to do is to evacuate the area and call the emergency services.

Oxygen

Oxygen cylinders contain the gas at a much higher pressure, as it does not liquefy at normal temperatures and a full cylinder will possibly be at a pressure of 3,500lb per square inch, or 230bar. This obviously means that the cylinders designed to contain this sort of pressure will need to be very heavily constructed. Handling of these cylinders will need careful consideration due to their weight and of course work boots with toe protectors will need to be worn. Oxygen regulators and fittings will have normal right-hand threads so as not to be confused with fuel gas fittings. Although oxygen aids normal combustion, but it is not flammable in itself. In fact, most forms of combustion will only take place in the presence of oxygen, or an oxidizing agent (an oxide-rich compound). All combustion

---

CYLINDER SAFETY

- ◆ Never use cylinders for rollers.
- ◆ Never bang or drop cylinders.
- ◆ Always remove regulators before moving cylinders.
- ◆ Ensure cylinders are restrained against a wall or in a trolley before use.
- ◆ Always use in an upright position.
- ◆ Store cylinders outside.
- ◆ Keep cylinders away from heat sources.
- * Propane and butane are heavier than air, so do not store in a hollow or cellar.

---

is a form of chemical reaction and under normal circumstances it requires the input of heat to initiate the reaction, that is, burn, but pure oxygen will react at any temperature with certain materials.

The biggest danger comes from oil and grease, which will spontaneously combust in the presence of oxygen. Therefore it is imperative **never to use oil or grease on any fittings** in gas heating equipment. It follows that oil and grease on overalls could pose a danger, so never use the gas cylinders on the trolley as a clothes rack to hang overalls on or over, as a small oxygen leak could spontaneously ignite, with disastrous consequences right on top of the gas cylinders. The same logic dictates that oxygen should not be used as a substitute for compressed air.

HEALTH AND SAFETY

Today, everything is under the full scrutiny of health and safety, rightly so in many respects, but in an industrial environment this is all taken care of and you are told what you should and should not do without any input from the individual. However, in your own home workshop, it is entirely your responsibility. Whenever any task is to be undertaken in a working environment, the usual practice is to write a risk assessment, pointing out any risks associated with the task and what steps need to be taken to eliminate or at least reduce the risks. When trying any soldering operation, especially for the first time, it may be a good idea to run through the whole process on paper, making a note of any potential risks that you can think of. At least then you will be aware of them should they become apparent during the process.

It could aid the process if a 'dry run' is undertaken, with everything in place in the brazing hearth except the lit blowlamp. In this way, you will see exactly where the flames from the blowlamp will be pointing and so on. This will be more relevant when silver soldering or brazing due mainly to the increase in the temperatures required; obviously more thought will be needed regarding your own personal protection and of course the increased fire risk. It may be prudent to run through what would happen if a fire were to occur, so as to know where the fire extinguisher is situated and how it works. In the event of a real emergency, this would save time; maybe only seconds, but this could be enough time to prevent the fire spreading.

Propane and butane are heavier than the surrounding air when released from the cylinder, which has health and safety implications. The gas will sink downwards and may collect at the lowest point it can find. In certain circumstances it will flow along the ground, just like water does; if it flows into drains, it can travel a long way from the original source of escaping gas, collecting dangerously at the lowest point. If this gas finds an ignition source, it may catch fire and cause severe devastation if an explosion ensues. This is why recommendations for the storage of these gases necessarily dictate that storage should be outside, preferably in a purpose-made, locked cage with the correct warning signs on display. A general-purpose workshop may have a pit built into the floor to facilitate access for vehicle maintenance. This could lead to a disaster if gas was allowed to accumulate down in it, as all it would take to ignite the

gas would be a spark from a grinder or a spanner hitting the floor, or even brazing operations in a brazing hearth. Propane and butane are both odourless gases, so to help detect a leak an odorizing agent, known as a stenchant, is added to the gas in the cylinders.

Fires and Extinguishers

The recommended method of tackling a fire of gas equipment is to, if possible without injury, cut the supply of gas to the fire, preferably shutting off the valve on the cylinder. This will allow the burning gas to subside and hopefully extinguish itself. If the fire continues burning once the gas source has been cut off, this can be tackled if it is safe to do so with a suitable fire extinguisher. If it is a large fire do not hesitate, call the fire brigade immediately.

THE ROAD TRAFFIC ACT

When carrying compressed gases in a vehicle, you will be obliged to follow the

When carrying compressed and inflammable gases in a vehicle, it is the driver's responsibility to affix the correct signs on the vehicle in order to comply with the Road Traffic Act.

CLASSES OF FIRE AND FIRE EXTINGUISHERS

Fire extinguishers are designed to be used on three main categories of fires, depending upon the type of material that is burning:
- Class A – Fires involving paper, cardboard, wood and so on.
- Class B – Fires involving flammable liquids, such as petrol and so on.
- Class C – Fires involving flammable gases.

Before any heating takes place, ensure that you have fire extinguishers to hand and that you know how to use them, in case of an emergency.

A fire blanket can be useful for smothering a small fire.

Road Traffic Act, which requires that the correct signage is prominently displayed on the vehicle. This is a diamond shape with a depiction of a flame for flammable gases, with the words 'Flammable Gas' across the bottom; for non-flammable gases such as oxygen, the sign required is a green diamond with the silhouette picture of a gas cylinder, with the words beneath 'Compressed Gas'. In the event of an accident the rescue authorities would then know what is being carried in the vehicle and can take appropriate action. The driver of the vehicle is responsible for the correct signage being displayed on the vehicle.

PERSONAL PROTECTION EQUIPMENT

As with all processes, the main list of essential tools is easily identified, but with today's emphasis on safety this needs to have careful consideration before any work is undertaken. The most obvious danger to the operator when soldering is heat, so bodily protection will be a must, with overalls being at the top of the list. Consideration also has to be given to the corrosive nature of some of the acid fluxes that may be used and the detrimental effects on the eyes and skin, as there can be a lot of spitting and spluttering, especially when heating up from cold.

Once things have been heated to silver soldering or brazing temperature, it will take quite a while for them to cool down to anywhere near a temperature to handle. This time will be extended if the items are left in the brazing hearth, as a lot of the surplus heat generated during the job will inevitably have been absorbed by the firebricks within the hearth. When the items are still red hot it is obvious that they will burn, but once they have cooled down to below red heat it is no longer so clear, but they will still be at several hundred degrees and will cause severe burns if touched.

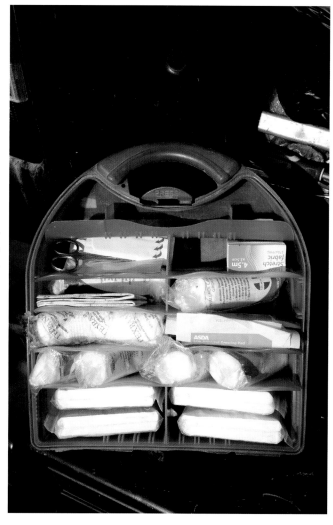

A well-stocked first-aid kit is essential for any minor accidents, even in the home workshop.

First Aid

All workshops, whatever the size, should have a well-stocked first-aid kit, as it is likely that at some point while in the workshop you will sustain a cut, graze or burn, however careful you are.

Burns, Cuts and Grazes

With any burn the quicker that you can take the heat out of it and the surrounding area the less it will hurt, so if possible either hold the burnt area in a bowl of clean cold water or under a running cold tap until the pain eases. If the burn is bad, cover it with a lint-free bandage and seek medical help as soon as possible; do not put any creams or potions on it, or listen to any old wives' tales about covering it with butter.

If you cut yourself and it is only an insignificant nick or graze, make sure it is clean and stick a plaster on it to protect it, as you are working in a dirty environment, however clean your workshop may appear. If it is more serious with lots of blood, the main priority above all else is to stop or slow the bleeding and seek medical assistance. It is probably a good idea to get someone else to drive you to the Accident and Emergency Department at your local hospital or to your doctor's surgery, or phone for an ambulance as you may at this point be in shock.

Learning by Experience

Once you have burnt yourself on something rather hot it somewhat sharpens your awareness to such dangers. I personally learnt this danger while in the school metalwork class many years ago. Although not silver soldering or brazing at the time, I was making a candlestick, with the part that received the bottom of the candle being heated in the forge to red heat to flare it out with a mandrel. This entailed holding the piece of metal in the forge with tongs so as not to lose it in the fire. Once to red heat, it was placed on the anvil and a pre-shaped mandrel was knocked into it with a hammer. This all went very smoothly, but I had carelessly put the extremely hot tongs on the floor. The next step was to quench the newly flared candle holder in the water bosh, so I picked up the tongs from the floor, but without paying attention to which end I had got hold of. Although the tongs were not red hot, they were at a black heat of several hundred degrees and the result was that the skin on my hand dried instantly. Despite dropping them quickly, it was too late and the damage was done. Luckily for me, the burn was only superficial with no lasting damage, but I am still careful many years later to check the temperature before I pick any hot object up.

Goggles

Goggles and safety glasses are readily available from most DIY outlets, tool shops and suppliers of soldering equipment. Even while soft soldering at relatively low temperatures it will be a good idea to wear safety glasses when using acid flux, because as it is heated it can spit and splutter, however careful you are while using it. When working at the higher temperatures required for silver soldering and brazing operations, there is a greater risk from things spitting while heating and the molten alloys with their higher temperatures will give nasty

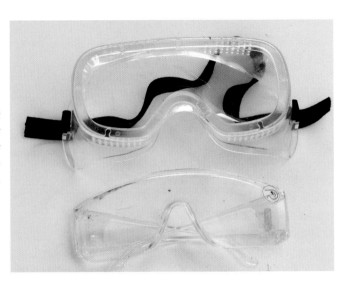

Goggles are important for eye protection. It is not only metal particles that they keep out, but also sharp shards of set flux.

The full-face mask protects the whole face as well as the eyes from flying debris.

burns, particularly to the eyes and surrounding area. When using an angle grinder for preparation and subsequent cleaning operations, it is imperative that safety glasses or goggles are worn, as with the grinder revolving at over 10,000rpm, particles will be flying at high speed all over the place.

Welding Mask

Although during silver soldering and brazing operations eye protection is required, it is only against flying detritus. When welding processes are described in the next chapter during the construction of a brazing hearth,

eye and full-body protection will be required against the extreme ultraviolet and infrared radiation from electric welding. A full-face welding mask, with the correct grade of filter for the amperage of welding that is being undertaken, as well as covering the body, will be required.

Gloves

With the nature of the processes involved during soldering and brazing, the operator's hands will be in the direct line of fire while working. The best type of gloves to wear will be welding gauntlets, which will afford

When welding, the hands are not only vunerable to hot sparks; the ultraviolet light produced when electric welding will also burn. Wear appropriate gauntlets.

When undertaking any sort of welding it is imperative the right welding mask is worn to protect the operator's eyes.

Before any soldering or brazing operations, don overalls and gloves suitable for the task. Stout boots should be worn as well to protect the feet from falling objects.

some protection to the operator's forearms as well, although these may be deemed rather heavy and cumbersome. An alternative would be to use a pair of gauntlets specifically designed for Tungsten Inert Gas (TIG) welding, where the operator requires manual dexterity while working, feeding in the filler rod, similar to soldering and brazing. If nothing else is available in the workshop other than a pair of general-purpose work gloves, these will afford at least a minimum of protection.

Overalls and Boots

Of course, it will not only be the operator's hands and eyes that will require protection. It is good workshop practice to wear overalls and stout boots while working, whatever the task. As with welding, there is a risk of overalls burning, so at a minimum cotton overalls should be worn, not nylon ones, which would melt and stick to the skin,

The leather apron is ideal to wear when using a brazing hearth for any length of time, as it gives the operator some protection from the heat being produced.

When brazing, the hair is vulnerable to stray sparks and excess heat from the blowlamp; stylish bandanas and hats made from flame-proofed materials are available at good welding outlets.

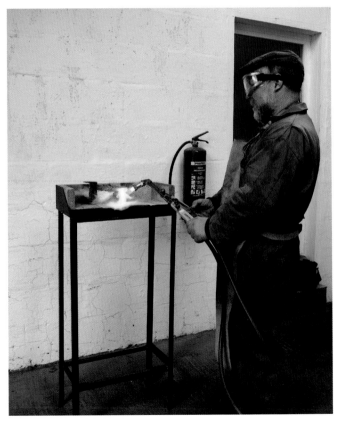

When the brazing hearth is in action, lots of heat will be produced, so sensible precautions need to be taken to avoid any accidents (for example, a clear space around the hearth and fire extinguisher to hand).

giving some nasty burns. Treated welding overalls, which will not burn, are ideal for soldering and brazing and are available at reasonable prices from welding suppliers and good tool shops. It is also a good idea to wear a leather apron while working at the brazing hearth, the sort traditionally worn by the village blacksmith, as the leather is somewhat self-extinguishing if caught by stray sparks and in use with the brazing hearth will afford some extra bodily protection from the radiant heat being produced.

ANCILLARY EQUIPMENT

With any operation, there are always tools and equipment required that are secondary to the main procedures, but nonetheless are just as important. The main requirement for tools in soldering and brazing is in the preparation of the materials before joining; these are mainly for cutting and cleaning so that the surfaces to be joined are adequately free from surface oxides before heating and are discussed more fully in Chapter 6.

Clamps

Most items to be soldered or brazed will require holding together while the heating operation is being carried out. There is a plethora of clamping devices available, from the common G clamp through to overcentre locking types such as 'mole' grips. As will be discussed later, the idea when soldering or brazing is to get the filler to flow by capillary action between the items being joined, so if clamped too tightly there will be no gap in which the alloy can flow, leading to a defective joint.

Jigs

If several similar operations are to take place within a short space of time it may be worth your while to make a jig to hold the various parts in correct alignment, especially if they

have an irregular shape or are awkward to hold in any other way. This is where the workshop scrap bin becomes invaluable, as items and offcuts that have been discarded may be just right to adapt to make a jig. That is why you should never empty the scrap bin too regularly!

Wire

A traditional way to hold things in place is to use soft iron wire, which can be readily purchased cheaply on a roll and is easily manipulated and cut with nothing more than a pair of pliers. Being of a soft nature it can be twisted together with the pliers, tightening a loop around two items until the correct tension is applied. Once the joining operation has been completed, it only takes the minimum of effort to remove the wire. There is no reason why, for soft soldering, copper wire could not be used instead, but for silver soldering or brazing the copper wire would probably melt just as brazing temperature is reached in the heat of the flame.

A roll of iron wire – this is useful for holding components together while silver soldering or brazing.

Various clamps – when soldering or brazing, these are always handy for holding things in place.

3 Designing and Building a Brazing Hearth

If any amount of silver soldering or brazing is to be undertaken, some form of brazing hearth will be a necessity rather than a luxury. The use of a blowlamp and the need to heat up the whole item before the joint is made with the filler rod means that a lot of heat has to be put into the job. The heat transfer from a gas flame is less efficient than any form of electric welding. The consequences of this are that a lot of extraneous heat will be all around the working area, which could have serious consequences regarding potential workshop fires.

With a single point heat source, such as a gas blowlamp, a lot of heat will be wasted. Apart from the safety aspect of using a brazing hearth, it also serves to absorb some of the heat that misses the item being soldered. This has two advantages. The first is a more even spread of heat reaching the item, as the surplus is absorbed by the firebrick of the brazing hearth and then radiated back towards the item, helping to heat it up quicker. The other is that less heat will be required, with the possibility of a smaller nozzle being used on the blowlamp, which of course will save gas and money.

DESIGN IDEAS

As noted earlier, brazing temperatures can be in excess of 950°C. Materials used to build a brazing hearth therefore must be able to

When building a brazing hearth it must be remembered that any materials used should be able to withstand high sustained temperatures.

withstand these temperatures, sometimes over prolonged periods. In this design, standard-sized firebricks will be used, as they are relatively cheap and are available in various sizes, which can be selected to suit the design that is used. Although a complete design is given here, it is only one idea and can either be followed or adapted to your own individual requirements. Alternatively, commercially made hearths are widely available, although of course there is no fun in this and buying one will simply deplete the learning curve of practical things and ideas in the workshop, which is the essence of this book.

Under no circumstances think that firebricks can be substituted with ordinary household bricks. This could be very dangerous indeed, as, although the bricks may

OPPOSITE PAGE:

The brazing hearth – a simple object designed to keep the heat where it is needed, around the brazing or silver soldering job.

A selection of angle and flat iron strips make an ideal resource for building your brazing hearth.

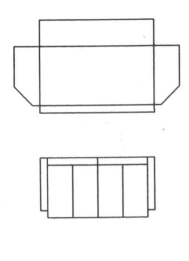

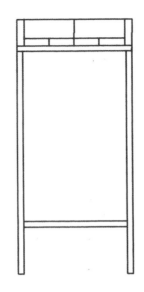

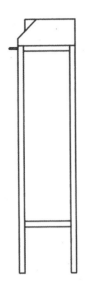

An outline plan of a brazing hearth; no dimensions are given, as these depend on the firebricks used.

look and feel dry, under the surface they may contain moisture and upon heating with the intense flame of a blowlamp, this moisture will soon turn to steam. The consequence of this is that the steam will build up pressure within the surface of the brick and pieces of its surface will then fly off with great velocity, in all and any direction. If any pieces hit your body it will hurt, but you could possibly lose an eye if you were unlucky enough to have a piece of hot brick hit you there – although of course you should be wearing safety glasses. But, even so, the use of anything other than dry firebricks for hearth construction could endanger innocent bystanders or even the pet dog or cat.

The most obvious criteria for the design of a brazing hearth is that the components need to be heat-proof or non-flammable; for example, the copious volumes of heat used in a brazing operation would soon have a wooden frame smouldering. The best material for any frame, or indeed a solid tray, would be steel. Stainless steel would make an admirable brazing hearth, but if working to a tight budget it might be possible to source cheaper offcuts. Mild steel would be the best choice, in angle iron, strip or sheet form, as it will withstand the heat being produced and of course it is extremely durable. The only downside to using mild steel is

that it will rust if exposed to the elements, but it could be given a coat or two of heat-resistant paint. This will of course enhance the finished results, but will have no effect on its use.

The choice of angle iron/steel strip construction versus sheet material is down to personal preferences and your own skill levels, before any consideration is given to what workshop equipment is available for construction. If a minimum of tools is available, it will be possible to construct a hearth using nothing more than a hand drill along with bolts or rivets for construction. If more sophisticated machinery is at your disposal, such as welding equipment, this opens up other avenues for building. My own preference is to construct a hearth using welding as I have the equipment already in the workshop and it is quickly done, with very little finishing work to do other than a lick of heat-resistant paint. A hearth from folded sheet could be made, although a suitable folding machine large enough to handle the envisaged size would be needed. Folding can be done with some sturdy angle iron and various clamps, but this will all take time and it is the finished article we want, in the minimum time, in order to be able to start doing some brazing.

Another point worth considering at this

stage is whether the brazing hearth is to be free-standing or bench-mounted. This is really a personal choice and is probably dependent upon how much space you have available in the workshop and how often you will be using the hearth. Like a lot of things in life, you may not know how much use you will have for the hearth until after you have built it. For example, if you have a particular project in mind, such as a locomotive boiler, the brazing hearth may be built to the required size for this project, but may never be used again once construction is finished.

Firebricks

Perusing online suppliers of firebricks it soon becomes apparent that they are available in a multitude of sizes and thickness, with corresponding prices to match. Sometimes suppliers may have a clearance corner where a limited number of bricks is available, which may be just right for your brazing hearth and you can make everything else fit around these for a bargain price.

There are two main types of firebrick readily available. The most durable type is the one made from fireclay, which will withstand temperatures far in excess of what we

Standard firebricks are ideal for the base of a brazing hearth. Although they do not reflect as much heat as other types, they are more robust.

A vermiculite block, which is very good at reflecting heat, but is not as robust as the ordinary firebrick.

will be using. The alternative is the firebrick made from vermiculite. The benefit with this type of brick is that the vermiculite has a tendency to reflect the heat back, without it being absorbed as much as happens with the fireclay type, and it can be cut and shaped by DIY tools fairly easily.

Possible Alternatives

Internet forums on the subject of brazing hearths abound with suggestions of alternative materials for hearth construction. One such suggestion is the heat-storing bricks that are used within electrical storage heaters. At one time, a pile of these bricks and the metal carcass could be seen on many a doorstep where they had been thrown out due to being too expensive to run and could be had for the price of asking. While these bricks will undoubtedly withstand the heat used for brazing, they are made specifically to store heat, which is the reverse of what we are looking for. At a push they may suffice for silver soldering at the lower end of the temperature scale. I have used one of these bricks for a small job of silver soldering some reducing sleeves into

the valve spring seats of a cylinder head with a small gas blowtorch; it did the job, but only just.

Another suggestion is to use the offcuts from building sites of extremely light aerated building blocks. This would probably work, but to my mind is a little foolhardy – these blocks are designed for building walls and their strength lies in holding the building together, not withstanding heat of many hundreds of degrees. They possess insulating properties due to the fact that they are made by an aeration process as the blocks are moulded and are full of tiny air bubbles. The aggregate is a mixture of gypsum and fly ash from gas-fuelled power stations so should withstand the heat; it is the binder that would fail.

The downside to this free bounty is that there is no telling what the blocks or bricks might have been in contact with, especially if found in a building-site skip. There are myriad chemicals that may have soaked into them, which could be detrimental to your own health, let alone the soldering or brazing job being undertaken, as these chemicals are vaporized by the heating torch. If the bricks or blocks have been stored outside it is fair to say that they will have absorbed water from our inclement weather. There is a danger that on heating this will turn to steam, sometimes with an explosive force, blasting bits of brick around the workshop. As mentioned earlier with regard to house bricks, the least effect this will have is that the steam produced will have an adverse effect on the work.

Reflective Blanket

A reflective blanket made from kaolin wool or similar heat-resistant fibre is available at a reasonable price and can be placed around the items being brazed within the brazing hearth. It will reflect surplus heat back towards the job, which will not only speed up the heating process, but will mean that less heat will need to be input for the

A nest of broken firebricks or vermiculite allows the heat from the torch to reach around the job.

moved too fast, the weld will be thin and patchy; too slow, and there will be an excess build-up of weld, with the possible consequence of blowing holes in the welding as too much heat builds up in one place. As with most things in life, practice makes perfect, and MIG welding is no different. A practice session on some material the same thickness as that to be used for the project is a very good idea. Try out some of the welds that will be required, so that variations of wire speed and voltage settings can be observed, until you are happy with the welds that you are producing.

Although gasless MIG welders, which use a special flux cored welding wire, are now available, welds made using this process will need to be cleaned after welding to remove the slag left on the surface from

job. Reflective blanket can easily be cut for custom applications.

WELDING

Using welding techniques will make a speedy and sturdy job of your brazing hearth, but unless you are proficient in any welding discipline, this skill will need to be learnt before proceeding with construction. Although there are many ways of welding that could be used for a project such as this and there are many good books available on the subject, Richard Lofting, *Crowood Metalworking Guides – Welding*, as noted above covers all common forms of welding suitable for the home workshop.

The simplest to master is Metal Inert Gas (MIG) welding. Once the equipment is set up and ready for the size of material to be used, it will only take a few practice runs before enough proficiency is gained to be confident enough to undertake the welding required for this project. All welding for this project will therefore be demonstrated with the MIG welder.

MIG Welding

If purchasing a new welder, it will come with rudimentary instructions for setting up the equipment, along with settings for wire speed, current and so on for the material thickness to be welded. If everything

is set up correctly, on pulling the trigger and starting the wire flow, it just requires the torch to be moved along the line to be welded at a steady speed, much like drawing a line with a pencil. If the torch is

The MIG welder set-up – an ideal welder for building your brazing hearth.

Practice welding runs on the thickness of material to be used will help to get the right settings. Left side – current too high; middle welds – about right; right side – current too low.

A good weld on angle iron, this example being a butt weld with good penetration, as used on the legs of the project hearth.

the flux core, much as when arc welding. It will be a matter of personal choice as to what welding method is to be used on your project. It may be that someone you know has welding facilities that you could use to save the initial outlay of new equipment, either doing it yourself using their tools, or getting them to do the welding for you once you have the hearth all cut out, in exchange

Drilling, bolting or riveting as an alternative to welding can be used to construct your hearth.

for cash or the use of skills you might possess that they do not.

Obviously other tools will be required to build the brazing hearth, most of which should be available in a modestly stocked home workshop. The most logical plan would be to draw out your design with the sizes of firebricks that you intend to use, either to scale or at least with measurements. This will allow you to work out the lengths of materials that will be required and you will be able to draw up a suitable cutting plan so that there is very little waste of those materials.

TOOLS REQUIRED FOR THIS PROJECT

◆ tape measure or ruler
◆ square
◆ hacksaw
◆ files
◆ angle grinder
◆ welder, or alternatively a method of drilling holes and setting rivets

A good selection of clamps, hacksaw and so on will be invaluable when building a brazing hearth.

The ideal design would be to have a completely self-contained unit, possibly on casters or wheels, so that brazing operations could be carried out in the fresh air outside of the workshop. This set-up may not be possible if your circumstances do not allow the amount of room required for this. A compromise would be to make a brazing hearth that can stand on top of a workbench. Although this would not be an ideal solution, so long as all health and safety precautions are followed, it should be safe enough and suffice for small brazing jobs. Although the firebricks will absorb the heat from the brazing, it would not be a good idea to place them in direct contact with a wooden workbench. The way around this would be to build a bench top unit with some short legs, which would give some clearance between the firebricks and the bench. As an extra precaution, a sheet of steel could be placed on the bench before the brazing hearth is placed on top. As already stated, when in use the brazing hearth must be in an area where there is adequate ventilation.

CHOICE OF MATERIALS

There are no hard and fast rules for the choice of materials to make the brazing hearth frame. The most obvious answer is to use mild steel angle iron for the main frame and legs, with flat strip used elsewhere that support is required. If a version with legs is envisaged, then with a few extra pieces of angle iron a very useful shelf could be incorporated into your design, which would add some rigidity to the whole assembly, especially when taking into account the weight not only of the component parts of the brazing hearth itself, but anything else that will be placed on it.

FINISH

Although not strictly a necessity, a painted finish on the steel work of the brazing hearth would make it look rather smart. Just the

The built unit, ready for a coat of paint and its firebricks.

steel work without any finish added would last for many years, but the rusty crust that would inevitably build up would soon start to look shabby. As the steel work is near to heat sources a coat or two of a heat-proof paint would be ideal. There are many types

available, but possibly the ideal choice would be one of the heat-proof paints that are available for the coating or recoating of heating stoves and which usually give a satin or matt finish in black. Once coated, to achieve maximum durability the paint will need to be heat-cured. Obviously when painted on to a stove it soon cures with the heat from the stove, but in our case it would be better to heat the framework up with a blowlamp, before placing the firebricks into it, in order to cure the paint, as most of the frame will not get enough heat under the firebricks to start the curing process.

ACCESSORIES

It is possible that while in the middle of a silver soldering or brazing operation you may want to put the flaming torch down for a minute or two, or have a pair of tongs readily available for use as the work proceeds. This is easily remedied by the fitting of a rail along the front of the hearth. This can be made from a length of 8mm or 10mm diameter iron bar, or whatever you have in stock that is suitable. For simplic-

Before placing the firebricks in the hearth, a coat of heat-resistant paint was applied, giving the unit some resistance to corrosion.

ity, it can be welded directly to the frame of the hearth, although it would be preferable to make brackets on the hearth to which the rail can be bolted. This would facilitate future modifications should the occasion arise, without having to resort to grinding and so on to make the changes.

If the brazing hearth is to be used inside it is a good idea to make a hood out of sheet steel to which an extraction fan can be connected, vented to the outside. For the air to go out through an extraction fan, an air vent will need to be installed to let the same amount of air back in, or the door or window in the workshop could be left open while working. The extraction fan that is fitted will need to able to withstand heat, as the rising air from the brazing operation will be hot

The finished brazing hearth set up ready for use.

USING YOUR BRAZING HEARTH

The obvious comment is 'Place the items to be joined in the hearth and light the torch', which in essence is all that is required, but a few 'dos and don'ts' will make the experience better for you. The brazing hearth may only be used for an extremely small percentage of your workshop time, and as most of us have not got enough space in the workshop it would be to easy to use the hearth as just another work surface or somewhere to stand things. Please resist this temptation. As careful as you may be, it is inevitable that at some point things will get spilt on the firebricks, contaminating them over time with an undefined cocktail of workshop chemicals. Even though these will be wiped up, a residue will have already soaked into the surface, waiting to be vaporized on lighting the torch. It may be a good idea to make a cover for the brazing hearth so that, once cooled down, it can be covered up. This will eliminate any problems due to accidental spillage.

Do not be tempted to use the brazing hearth for soft soldering operations. It will do a magnificent job of containing the heat, but contamination from the flux and soft

solder will affect the quality of later silver soldering or brazing done in the hearth.

The theory and statistics of which torch is powerful enough for whatever job have been discussed above, but from a practical perspective there are various things that will

rear their respective heads as you progress with your brazing. When working inside a closed object, such as a model steam boiler, when silver soldering in the stays in the firebox, for example, it will be found that the flame on an ordinary gas torch will keep going out. In these situations, it would be advantageous to use a cyclone nozzle on your torch. This type of nozzle draws the air necessary for combustion of the gas from the opposite end of the nozzle to which the flame issues, mixing the two together and thus keeping the flame alight in confined spaces.

Experience will soon dictate how you can proceed with placement within the hearth and where to put reflective blankets and spare firebricks. It will be a good idea to keep any pieces of broken firebrick for this purpose, so that if necessary they can be packed around the items in the brazing hearth. This has a twofold effect: first, it will help to hold the items being soldered; and, second, it will help to reflect heat back into the job as it proceeds.

The brazing hearth in use. Note the clear surroundings, free from clutter that could constitute a fire hazard during and after use.

After much consideration, I decided that a tray made from sheet steel would be the best solution to hold the firebricks, as they are fairly fragile in nature and if they cracked during use, the tray would contain the broken pieces. I used some pieces of angle iron and clamps to create the folds in the sheet and used the MIG welder to join the seams. The legs were made from angle iron, which was also welded before it was painted.

1. After careful measurements were taken from the firebricks used for the project, I marked out the dimensions on to a sheet of steel. The corners of the folds were drilled before cutting out. This gives relief when folding the sides up and also makes cutting out easier with snips or a bench-mounted shear.

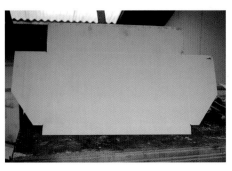

2. Here, the sheet had been drilled and cut ready for the folding operation. This was done with clamps and mild steel bar, before using a panel-beating hammer to produce a nice tight fold.

3. The cut corner can be seen here, showing the relief obtained by drilling a hole in the corner. If the folds are made without this relief, a puckered corner will result; this will still be functional, but not very attractive.

4. It is important to remove burrs from all the edges that have been produced by the cutting process. I used a file, but an angle grinder with a sanding disc fitted could have been used. Gloves should be worn to avoid cuts to the hands.

5. Folding the long sides was done by clamping the sheet between two long lengths of steel, making sure that they were exactly on the fold line. The third and fourth bends required a short length of angle to be cut to fit inside the tray, before the folds were made.

6. To produce a clean fold using a hammer, work along the fold line in stages, using a glancing blow in the direction of the fold. Once the fold has been made, tighten the corner by working along it with the hammer against the steel below. Do not hit too hard, otherwise the metal will stretch.

7. When I was happy with the folds, the corners were tack-welded before seam welding, then the welds were dressed back with an angle grinder fitted with a sanding disc.

8. A trial fit of the firebricks was made at this point, before the legs were made and the whole unit painted. At last, the finished item was looming and would soon be ready to use.

9. I decided that the simplest way to make the legs was from some 25 × 25 × 3mm angle iron that I had in stock. After marking out carefully, the parts were cut with an angle grinder fitted with a cutting disc to size.

10. The parts were clamped so that they were square to one another and then MIG-welded together, forming four stout legs. These were then welded to the tray for simplicity. They could just as well have been bolted on, but this would have caused a problem with bolt heads under the firebricks.

11. I had made a rail for the front of the hearth so that the torch could be put down or used for holding tongs and so on. I fitted it at this point and realized it was a little low, but still usable.

12. After giving the whole unit a coat of heat-resistant paint, all that was left to do was to place the firebricks into the tray and give the hearth a test run to help cure the paint.

4 Brazing

Brazing is carried out at temperatures approaching the fusion temperatures used in welding. Although brazing is not as strong as fusion welding, it is the strongest form of metal bonding without melting the parent metal of the components being joined together. Brazing will therefore require more heat input than any of the other soldering operations. Although the melting temperatures of the base materials will not be reached, they will not be far short. Whereas welding uses a concentrated source of heat, confining it to a relatively small area adjacent to the weld being formed, brazing uses a more general heat source in the form of a gas blowlamp. As with gas welding, the heat transfer is fairly inefficient, so much more heat has to be put into the objects being

Brazing rods are available in various sizes and alloys, with prefluxed and plain rods giving the operator the choice for the particular job in hand.

brazed than is actually being used to carry out the joining process.

Traditionally, brazing was carried out using spelter, or brazing rods made of brass, but today all sorts of metals are used in the brazing rod alloy content. These are predominantly of copper, with tin, nickel and silver, to name just a few other alloying metals that will be found in the modern brazing rod. For obvious reasons, rods containing the more exotic metals will be expensive and therefore reserved for jobs that cannot be done using any other metal.

JOINTS

Joint Overlap

Silver soldering and brazing operations require a lapped joint in order to be successful and strong. This overlap will allow somewhere for the filler material to be drawn into by the process of capillary action and it is the molecular bond between the parent metal and the brazing material that gives the joint its strength. The ideal overlap is three to four times the thickness of the sheet thickness from which the components are made – for example, if 1mm sheet is used, a 3–4mm overlap will be ideal.

OPPOSITE PAGE:
Brazing is a good, strong way to join parts together. One advantage is that it can be used to join dissimilar metals.

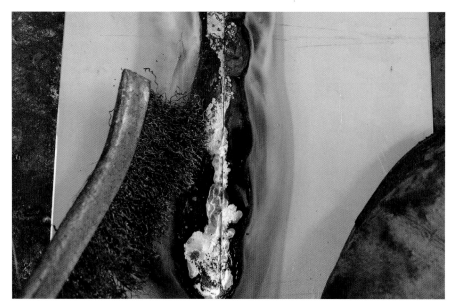

A lapped brazed joint on sheet steel. Better application of the flux would have resulted in a neater joint, but nonetheless a strong joint was made.

Joint Gap

The other important feature of creating a sound joint is the joint gap. For silver soldering or brazing alloy filler to flow, it needs a gap at alloy melting temperature. A tight-fitting joint will not allow any filler material to flow into it; equally, too large a gap will not work either. The actual gap required will depend upon the fluidity of the filler alloy used and will be listed in the data sheet for each particular alloy. Obviously, thinner flowing alloys will require a smaller gap, but a gap somewhere in the region of 2–8 thousands of an inch (0.05–0.15mm) will prove to be ideal. This gap needs to be maintained at brazing temperature and it must be remembered that dissimilar metals have differing coefficients of expansion, so as things are heated up movement will take place between the various items to be joined.

There are several ways in which the joint gap can be maintained. As mentioned in Chapter 1, alloys are available in strip form that are the right thickness for the ideal joint. Placing the foil strip into the joint before heating will set the correct gap, which will be maintained until the heat melts the foil strip. Fluxing will need to be carried out before heating to ensure a sound joint.

Another way to maintain the correct gap is to use a centre punch on one or both of the items to be joined. As the centre punch is forced into the metal by a blow from a hammer, the displaced metal forms a ridge around the centre punch mark, rather like the sides of a volcano rising above the surrounding ground. The harder the centre punch is hit, the larger the ridge. Careful use of the centre punch will produce a ridge to the correct height of the gap required. Some practice beforehand on similar materials will allow you to gauge how hard to hit the centre punch in order to obtain the correct size of ridge. If this technique is used, pay particular attention to getting the centre punch marks in the centre of the lapped joint so that when the joint is completed it is within the joint itself, leaving no indentations or marks externally for corrosion to

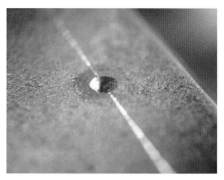

Close-up of a centre punch mark. The force of the blow from the hammer leaves a crater and a rim is formed from the displaced metal.

Using a hammer and centre punch on a marking-out line, a gentle tap will leave a light mark. Once you are sure of its position, use a heavier blow.

establish itself. On thin materials, striking the centre punch too hard will result in the point penetrating right through, if nothing else leaving an area of weakness. Also bear in mind when constructing a pressure vessel that any weaknesses within the materials used could cause catastrophic failure, for example, a centre punch mark from which a crack could originate.

Joint Integrity

From the outside, a completed joint may look very neat and sound, but appearances can be deceptive. For maximum strength the whole joint needs to be solid, with no gaps, voids or trapped flux. The only way to ensure this is in careful preparation of the items to be joined and the correct placement of the flux. The other important consideration is regarding how the heat is applied. It must be remembered that the capillary action draws the alloy towards the heat source, so think about the direction the heat is applied to the joint. Gaps or voids can be created by overheating of the filler alloy,

A cross section of a brazed joint, showing the integrity of the alloy within after being drawn into the joint gap by capillary action.

either by too high a temperature being used on the whole job, or possibly from localized overheating if the torch is held in one place to long, whereby one of the alloying metals boils off, leaving a dreaded void. As with all things in life, practice makes perfect and soldering and brazing is no exception. This will stand you in good stead for when you actually do some real brazing, as you will have more idea as to whether the joint is sound or not.

WORKSHOP SAFETY

In the previous chapter, the construction of the brazing hearth was shown in detail, as this is a 'must have' piece of equipment if any amount of brazing or silver soldering is to be undertaken in the workshop. With such large amounts of heat it is imperative to make certain that all flammable things are out of the way before lighting the gas burner, such as fuel cans, oily rags, cardboard, paper and so on. While concentrating on the brazing operation the last thing you want to do is set light to the workshop. With flammable liquids such as petrol, it is the vapour that is likely to ignite. It can travel quite a distance from the container and, once ignited, the flames will travel back to the container, with obvious disastrous results.

However, in a workshop environment there is a whole host of liquids that are a potential hazard, from cleaning materials

through to all the various thinners available for the differing painting systems of today. It goes without saying that you must have a suitable fire extinguisher to hand just in case and be aware of how to use it. Unless a suitable place inside the workshop can be set up with proper ventilation, it would be prudent to undertake brazing operations outside of the workshop. This does not decrease the fire risk associated with brazing, but it does save moving all those various bits and pieces that are flammable out of the workshop and at a stroke eliminates most of the ventilation issues.

When burning gas, usually propane, water vapour will be a by-product. Inside the workshop, if you have any machinery, such as a lathe or a milling machine made from cast iron, the warm moisture-laden air will deposit the moisture as condensation on the colder cast iron, with possible detrimental effects. When using acid fluxes while soft soldering, these will add to the corrosive properties of that moisture in the air. Poisonous fumes are another obvious concern, with metals such as cadmium causing breathing problems in the short term and cancer in the long term. Although alloys are now available that are free from cadmium, if using rods from an unknown source it would be prudent for to assume that they may contain amounts of cadmium and to eliminate the risk by being outside with plenty of air.

Close-up of a brazed joint. Once cleaned of flux residue, a neat finish will result.

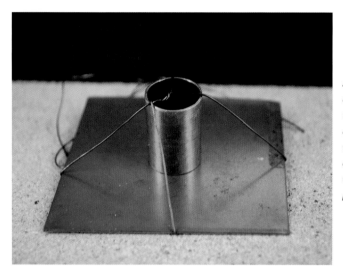

A good way to hold awkwardly shaped parts together during soldering and brazing operations is with soft iron wire, which can be twisted tightly with pliers, gripping the parts together.

Hard flux on a joint when cold looks like glass and is just as hard.

Plunging a brazed item into bucket of cold water has the effect of cracking the brittle flux.

THE BRAZING PROCESS

The first step is to clean the items to be brazed, removing all traces of oil, paint, rust and any other oxides. A method of holding the parts together needs to be thought out – do not forget that the parts will be red hot during the brazing operation. Large clamps will draw a lot of heat, which may well do damage to them, so something like soft iron wire is a good choice as it is easy to manipulate into position and to tighten by twisting with pliers and of course it will withstand the heat of the flame.

Once set up in the brazing hearth, allow space around the back of the object so that the heat can be manoeuvred all around for even heating. Use broken/small pieces of firebrick as props where necessary. The objects to be brazed can either be fluxed before heating by mixing the powdered brazing flux to a paste with a little water, or alternatively the brazing rod can be heated in the flame and the hot tip plunged into the tub of flux. Upon removal, the rod tip will have a quantity of flux adhered to it, which can be applied to the joint being brazed once the correct temperature is reached. This technique is fine for a small, quick job, but if it is a large job and the heating is over a longer period, the flux is best applied before heating, as this will prevent too much oxidation at the brazing junction.

Brazing is carried out in the 800–1,000°C range, so we are talking red hot. The flux will be seen to melt on to the hot surface and will take on the appearance of molten glass. This is the time to apply the rod to the joint, not into the flame. If the joint is hot enough, the rod will melt into the fluxed joint and be drawn along it by capillary action. Carry on heating and adding more rod until the joint is complete. If the rod is melted in the flame and the joint is not quite up to temperature, there is a risk of the joint failing as the rod material will possibly stick just to the surface of the objects being joined and not penetrate into the joint itself.

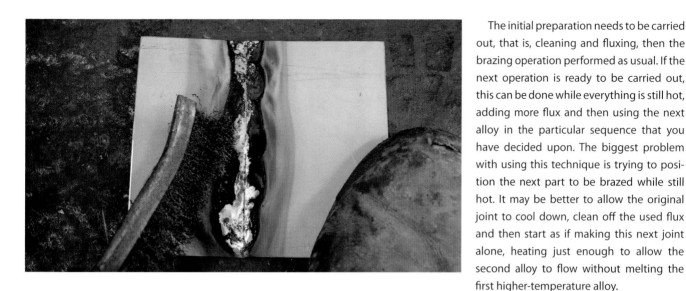

Once out of the bucket, the cracked and loosened flux can be removed more easily.

The initial preparation needs to be carried out, that is, cleaning and fluxing, then the brazing operation performed as usual. If the next operation is ready to be carried out, this can be done while everything is still hot, adding more flux and then using the next alloy in the particular sequence that you have decided upon. The biggest problem with using this technique is trying to position the next part to be brazed while still hot. It may be better to allow the original joint to cool down, clean off the used flux and then start as if making this next joint alone, heating just enough to allow the second alloy to flow without melting the first higher-temperature alloy.

One drawback of brazing is the flux residue that is left. Once cold, it will take on the appearance and texture of glass, which will require chipping off with a pointed tool. During this operation goggles or other eye protection must be worn, as when the glass-like substance is broken up, small shards will fly off in all directions, with very serious consequences if they get into your eyes. If the items being brazed are not sensitive to sudden cooling, for example, they are made from mild steel or copper, then as soon as the brazed item has cooled to below red heat and the braze has solidified, it can be plunged into a bucket of cold water. This will have the effect of rapid cooling and hopefully will shatter the solidified flux, making final cleaning easier and less time-consuming, but above all do not forget the safety goggles.

STEP OR SEQUENCE BRAZING

Step, or sequence, brazing is a good technique where a joint has been made and another one needs to be created. The technique is to use the higher-temperature melting alloy first, followed by the next lower-temperature alloy and so on until everything is brazed together. This technique works just as well with silver soldering.

After the first high-temperature brazing operation to repair a fuel tank take-off pipe, everything was cleaned before setting up for the second step.

The lower-temperature second joint was completed without melting the first joint.

MATERIAL SELECTION

While the material selection for whatever project you envisage is beyond the remit of this book, certain criteria and pitfalls are worth mentioning here, as they could lead to catastrophic circumstances if the wrong selection is made. For example, when building a boiler from copper for a model steam engine, it is imperative that the right grade of copper is used. What is known as oxygen-free copper will be required – either C103 or C106 grade of phosphorous deoxidized copper should be selected. Copper that is known as tough pitch, high-conductivity or electrolytic C101 copper will contain oxygen. On heating, this can produce steam, which breaks down and causes hydrogen embrittlement within the material's structure; it is this that will cause the structure failure. Also, with safety-critical work, careful selection of the filler rods will be vital, checking that there are not any metals present in the chosen rods, such as aluminum or titanium, that could alloy with the copper, forming an intermetallic compound and leading to possible joint failure at some time in the future.

ALUMINIUM BRAZING

Aluminium brazing is seen by some to be a mystical art best left alone, but it is in fact similar to any other brazing operation. The main difficulty is that aluminium and its alloys melt at relatively low temperatures and show no signs that they are about to melt, as other metals do by going red hot beforehand. The other problem associated with aluminium alloys is that they have a very tough oxide layer on the surface. This is what makes these alloys suitable for many applications, as once this oxide skin has formed very little further corrosion takes place. As we now know, it is the oxides that need to be removed before a successful soldering or brazing operation can take place. Pure aluminium melts at 660°C and its various alloys between 640°C and 655°C,

The traditional way to braze aluminium used a system of aluminium rods and a special acid flux to remove the stubborn oxide film.

whereas aluminium oxide does not melt until it reaches temperatures in excess of 2,000°C, although it does weaken somewhat at around 300°C. Either mechanical or chemical means will be required to cut through the oxide layer; due to its tenacity, very acid fluxes are required.

The Brazing Process for Aluminium

To perform traditional aluminium brazing with any success oxyacetylene equipment will need to be employed, as a small, hot flame is required. A large flame from a normal gas blowtorch would end up with the whole thing melted on the floor. Information regarding the use of oxyacetylene equipment can be found in Richard Lofting, *Crowood Metalworking Guides – Welding.* If you are experienced with gas welding equipment, the following will just need practice to perfect the technique; otherwise read it as an illustration of the technique.

Any cleaning should be done with a stainless steel wire brush, as an ordinary steel one will contaminate the aluminium surface. Also, any mechanical grinding should be done with a zirconium disc for the same reason before the brazing operation. It is actually brazing that is being done, so the aluminium should not be melted. The surface, once fluxed and heated, will start to glisten as the oxide film is broken down by the flux. Gently applying the filler rod, it should now melt into the surface. If joining two pieces of sheet together, keep moving the flame evenly in a swirling motion as with steel. Proceed swiftly along the line of the joint once the rod has started to melt so as not to melt through. Caution is needed as the end of the joint is approached, as the heat build-up will happen quickly, and you will end up with a pool of melted aluminium on the floor if not careful.

For use with gas brazing, a special flux is available that is mixed with water and applied to the parts to be joined and to the rod. Further flux can be added by dipping the rod in the flux mixture as you proceed. Once the work is finished, it is imperative to wash off any residual flux with hot water scrubbed thoroughly with a brush, as the flux is acidic and will eat away at the aluminium. This is the main disadvantage of using this type of brazing to join aluminium, as any work where the reverse side is out of reach and the flux cannot be removed, will rapidly corrode.

Unless you have bought the aluminium and it came with a specification, it will be nigh impossible to determine the alloy composition of the material. Various filler rods are available, from pure aluminium to those with varying amounts of silicone or magnesium in the alloy mix. The best choice for general-purpose work is a 5 per cent silicone rod, which will join most alloys. If a magnesium-bearing filler rod is used on aluminium that does not contain magnesium, there will be the possibility of cracks forming in the subsequent joint. As an alternative to the 5 per cent silicone rod, if there is spare material the same as being used to hand, a thin strip can be cut to use as a filler rod. This will ensure that the filler used matches the base material and will avoid compatibility issues, although this is actually welding.

THE MODERN WAY

There are now available alloys that are capable of joining most types of aluminium and its alloys. Some will successfully join zinc-based metals such as Mazak. This was a casting material used on classic cars for things like door handles, which were then chromium-plated, but it is very susceptible to corrosion. Alloys now produced by specialist companies, with trade names such as Lumiweld and AluBright, require very little heat input. This is due to the alloy mix lowering the melting temperature; for example, AluBright has a melting temperature of 380°C, so a normal blowlamp can be used for successful work.

As already stated, the biggest problem with aluminium and its alloys is the oxide film. Traditionally, a very aggressive flux was used to break it down, but with the new technique, once the items to be joined have been heated, the alloy is added to the joint area. The molten alloy sitting on the

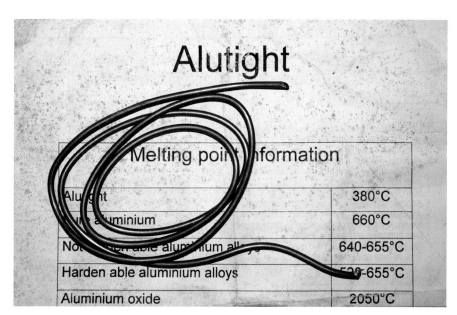

Melting point information	
Alutight	380°C
Pure aluminium	660°C
Not solderable aluminium alloys	640-655°C
Harden able aluminium alloys	538-655°C
Aluminium oxide	2050°C

The modern approach uses an alloy rod such as AluBright. These rods have a melting temperature of 380°C, but require no flux to braze aluminium and its alloys. There are other brands, such as Lumiweld, that do the same job.

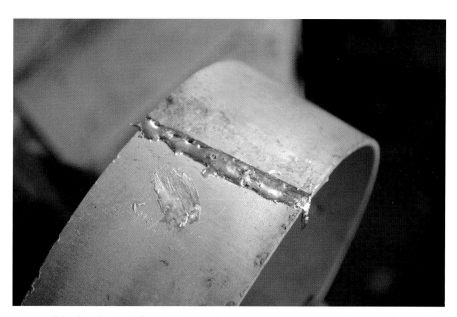

Using an Alubright rod is straightforward. Heat the aluminium with a propane torch until the rod adheres to the surface when rubbed over it, then scratch through the liquid pool with a screwdriver to break the oxide layer on the surface. Do not apply too much heat.

base material is held away from the metal's surface by the oxide layer. By the use of a stainless steel spike, the oxide layer is broken up through the molten filler alloy, whereby the filler alloy creates another alloy with the surface beneath the oxide. Any loose oxides float to the top, thus a joint made in this way is free from any acid fluxes to cause future problems with hidden corrosion and so on. It is claimed to have a tensile strength of 85N/mm².

There is now a second generation of these specialist rods that will penetrate through the oxide layer without the use of cleaning wires and the like, and will give a very strong bond. They are reputed to be suitable for repairing engine cylinder heads and sumps and so on, but as yet I have no 'hands on' experience with them.

BRONZE WELDING

Although this book is principally about the techniques of soldering and brazing, it would not be complete without mentioning the technique of bronze welding. In essence, brazing and bronze welding are one and the same thing, using the same heating apparatus, filler rods and flux. The difference is in the way that the filler rod is applied. As already discussed, brazing relies on the phenomenon of capillary action drawing the filler material into the joint being made. Bronze welding relies on a fillet of alloy being built up, although one is not exclusive to the other; for example, when brazing a holding stud on to an object, once the brazed joint has been made a fillet can then be built around the stud, adding more strength to counteract any shearing force put on to the stud in service.

In all probability in a non-critical job, it would suffice just to carry out the bronze welding, fluxing where the fillet is to be placed before heating. Once everything is at the melting temperature of the alloy being used, apply the rod to where the fillet is being placed, using the heat in the objects being bronze-welded, rather than the heating flame as with brazing. This will ensure that the temperature has been reached, rather than the filler just melting in the high-temperature flame. As the flux melts it will flow between the two components, cleaning away any oxides, then capillary action will draw in the filler without any effort by the operator as the fillet is being built up.

Again not strictly soldering or brazing, but nonetheless extremely useful in a repair workshop, the technique of building up a worn shaft using brazing/bronze welding rods must be mentioned. When a shaft has been running in a bearing for a long time, or has had scant servicing, sometimes the bearing lubricant will have dried out, or, worse still, been contaminated with grit or dirt. This will act as an effective grinding

The finished bronze welded joint showing a good fillet.

paste and will soon destroy the bearing surface. The usual occurrence, if the shaft is running in a bronze or brass bearing, is that because the surface of the bearing is relatively soft compared to the shaft, the grit will embed itself into the bearing surface. This creates a very effective abrasive, which then wears away the harder shaft.

The other most common scenario is that if the bearing is a roller or is needle-bearing, whether through lack of maintenance or servicing, or just that the bearing has gone beyond its service life, once the lubricant has all but expired and the surface may be contaminated with gritty detritus, the bearing will wear. There is a distinct possibility of the bearing seizing, the usual result being that the inner bearing track on the shaft, being only a push fit, will slip and rotate, which will cause rapid wear to the shaft. The bearing itself is easily replaced, but this will leave the shaft either requiring replacing or repairing. The repair could be done by welding, but this could easily lead to the shaft distorting due to the localized heating to fusion temperature of the shaft

Here, the preparation has been done to make a bronze welded joint with the two pieces held together with soft iron wire. This will prevent movement while heating and applying the flux and filler rod.

material. By adapting the bronze welding technique and using oxyacetylene equipment, the shaft can be built up and with the facilities of a lathe in the workshop, can easily be machined back to its original size, ready for the fitting of a new bearing. It is true that any heat applied to the shaft may cause distortion, but by using the gas torch sensibly the whole circumference of the shaft can be heated, lessening the possibility of distortion, and the bronze welding temperature is significantly below fusion temperature.

In my hobby of repairing and restoring vintage tractors, it is common to find that on the brake and clutch pedals, which are connected using clevis pins, the clevis pins often wear where they are in contact with the pedal or with the stirrup on the connecting rod. This can be effectively repaired by the bronze welding technique. True, the clevis pin could just be replaced with a new one or one turned on the lathe, but when built up with bronze, the characteristics of the bronze filler alloy are an advantage, as the bronze is resistant to wear and is easily worked, as well as being somewhat self-lubricating when run against steel.

BRAZING PITFALLS

- ◆ Avoid overheating the parent metal.
- ◆ Do not heat longer than necessary to avoid liquation of alloy.
- ◆ The use of certain materials can lead to hydrogen embrittlement.
- ◆ Cleaning melted flux can sometimes be a problem.

To illustrate the practical use of brazing, I have shown how to braze the hinge strengthener on a Land Rover door. Within the door channel there are two tubes that the hinge bolts pass through, which I rescued and reused after cleaning. As the job was done with the door skin *in situ*, I had to build the channel up from one side, leaving the two tubes to braze last of all. Two reasons that the job was brazed rather than welded were that it had been brazed originally and the tubes had traces of braze still on them, and that welding would have made the job difficult to clean up if any surplus weld entered the tubes.

1. Anyone who has owned a Land Rover of any model will know that the worst place for rot is the doors. After replacing the lower rail, the hinge strengthener was rebuilt using some of the original parts, with new ones being made up to replace the rotten or missing parts.

2. To effect a braze repair a heat source is required. I chose oxyacetylene, as it would keep the heat in the surrounding area to a minimum. The door skin is aluminium, so the less heat the better. The flame should be adjusted to slightly oxidizing, so as to prevent the zinc in the alloy boiling off.

3. Once the flame had been set, the ends of the brazing rod were heated, before plunging it into the flux. On withdrawal, a blob of flux was attached.

4. The parts to be joined were heated with the flame. Note when brazing with oxyacetylene the inner cone should not be used, as with gas welding, as it is too hot; use the outer larger cone.

5. The flame was kept moving so as to bring both parts up to temperature together. It is important that they are both hot enough, otherwise the braze will only adhere to the hot side.

6. Once both parts had an even glow, the flux-coated rod was touched on to the glowing surface. If hot enough, you will see the braze flow around and fill the gap.

7. Moving to the second tube to be brazed, the same procedure was followed. If you look closely it can be seen that the heating is uneven, with one side not up to temperature.

8. The heat source had to be kept moving until the whole joint was glowing, then the fluxed rod was added as before until the gap was filled.

9. The whole assembly was allowed to cool right down before proceeding to clean off the hardened flux.

10. It was necessary to chip off the hardened flux. I used general-purpose Sifbronze flux on this job. Wearing eye protection against the flying chips is a must.

11. A stiff wire brushing removed the last traces of the flux and any of the zinc coating that had oxidized from the Zintec-coated steel sheet used for this job.

12. At this stage, a coat of undercoat/primer was added for protection until the door received a full coat of paint, which would happen after the other door had been given the same treatment. Once the doors have been painted, the internal box sections will receive a wax treatment, which will hopefully keep the tin worm (rust) at bay for a few years.

Silver solder has many practical applications. Here, a bandsaw blade is awaiting heat from a blowlamp to melt the solder.

5 Silver Soldering

Silver or hard soldering can be seen as the middle ground between brazing and soft soldering. It produces a stronger joint than soft soldering, but not quite as strong as a brazed one. Nonetheless, it will be exceptionally tough, with the materials failing before the joint itself. This can quite easily be demonstrated by producing a test piece. Silver solder two pieces together as you would if constructing something, then once cooled down grip one edge in a bench-mounted vice and the other with a pair of mole grips or something similar. Keep wriggling and twisting until something breaks. If the joint has been made properly, it will invariably be the base material, not the joint, that fails.

The name, silver solder, is derived from the fact that traditionally silver solders contain a high proportion of silver in the alloy. With silver being one of the precious metals, this adds many qualities to the solder, although unfortunately this also includes the price. It has a very high resistance to corrosion and is free-flowing, so is an ideal constituent in an alloy intended for close-fitting capillary joints. It also has reasonably good tensile and shear strengths.

CADMIUM DANGERS

Cadmium, one of the so-called heavy metals, is a white metal with a melting temperature of 321°C. It was discovered during the early 20th century that by adding cadmium to silver solder, it imparted some very beneficial characteristics to the solder, such as good wettability and better mechanical properties. It also made the solder cheaper, as less silver was required in these alloys to achieve the same results. However, as with a lot of things today, what was at one time thought to be the best thing since sliced bread and did an extremely good job, has now turned out not to be any good for us and indeed the environment that we live in. The metal cadmium, along with many others, is one of these materials best avoided.

The dangers from cadmium oxide fumes are very real from silver solders bearing cadmium in the alloy mix. Cadmium is classed as a class 2 carcinogen and has a Maximum Exposure Level (MEL) to which anyone should be exposed in the working environment. The Health and Safety Executive (HSE) has developed a formula to work this out. It translates in real terms that in a correctly ventilated workshop working for eight hours a day, a maximum of 20g of cadmium-bearing solder per ten minutes can be used. This works out at approximately just over two rods of 2.5mm diameter every ten minutes. But, even so, unless the correct ventilation and fume-extraction equipment are in place, it is recommended that an alternative to cadmium-bearing solder is sought. There is a very real danger of damage to the lungs and kidneys in the short term, as well as long-term cumulative effects, with little indication until symptoms show themselves with possible cancers developing in future years.

It is not only the fumes while soldering that are a problem. Vaporized cadmium, distributed around the workshop by the effects of the heat from the heating torch, will settle in and around the workshop as the air cools down, leaving a film of cadmium oxide on all surfaces. This will be disturbed every time something is moved in the workshop. It is not only the breathing in of this oxide dust that is of concern; cadmium oxide is just as detrimental to health entering the body via the digestive tract. This is another good reason to wear protective gloves in the workshop environment, but, above all, make sure that you wash your hands thoroughly before eating or smoking, or indeed putting your hands anywhere near your mouth.

DECORATIVE SILVER SOLDERING

As noted earlier, silver solder is relatively expensive when compared to other solders and brazing filler rods. This is due to the cost of the silver content in the alloy, but when used for jewellery or decorative purposes where very little is actually consumed, the cost is relatively small in real terms. The course of events when joining small decorative items is to hold the pieces together with something like soft iron wire, which is easy to bend to shape but has enough strength to hold the items in place while being heated.

Silver soldering is used extensively in creating decorative jewellery; the art is to choose a solder that has a colour corresponding to the pieces being joined.

Two parts wired together ready to silver solder.

Flux should be placed only where the solder is to flow, possibly with a small paint brush. The solder under these circumstances is cut into short lengths and placed along the line to be joined at spaced intervals so that there is not too much solder – just enough to be drawn into the line by capillary action.

Once this has been set up in the brazing hearth, the whole ensemble can be heated with the blowlamp, moving the torch around to distribute the heat evenly around the items. As the solidus temperature is approached, the flux will start to bubble and will be seen to wet the surface where it has been applied and the solder pieces will start to melt. Continue to apply heat so that the liquidus phase of the solder is reached and all of the alloy will be molten. At this point, the solder will be seen to draw into the joint by capillary action, leaving very little solder visible at the surface. The result will be a strong, neat joint. All that will

now be required after turning off the gas is to clean the flux from the items and the job will be done.

It is imperative to heat up the parts to temperature as quickly as possible, as some of the alloys are subject to what is called liquation if heated too slowly. This has the effect of splitting the alloy, as the early melting constituencies of the alloy separate from the rest of the alloy before complete melting has taken place. Some soldering and brazing alloys are more susceptible than others to this phenomenon, but if carefully chosen by studying the manufacturers' literature, the chances of this happening can be reduced, if not eliminated, when combined with rapid heating up to melting temperature and completing the joint quickly.

When silver soldering is used for decorative work and jewellery making, or indeed repair, it may be more important to match the soldering alloy's colour to the materials being used, rather than focusing on the

ultimate strength of the joints. Joints made in silver on silverware and jewellery will be hard to detect when finished, cleaned and polished.

GENERAL SILVER SOLDERING

The most common approach to silver soldering a joint is the same as with brazing. The joint after cleaning is coated with flux where the solder is intended to flow, that is, in the joint. The pieces are then placed together, so that they cannot move throughout the operation; a good technique is to bind them together with soft iron wire. Once the job is complete, all that remains to do before cleaning is to remove the wire with a pair of pliers. The wire can either be saved for reuse or put in the scrap bin.

Manufacturers' Data Sheets

For decorative purposes, the selection of the soldering alloy has only to be of the required colour for the finished joint. Where items are being soldered to contain liquids or are under pressure, or both, other factors in choosing the correct filler alloy will need to be considered before the job is undertaken. This is where the soldering alloy manufacturers' data sheets are invaluable, as the constituent alloying elements are listed and in the percentage that they are present. Also, most give the ultimate tensile and shear strength that will be attained with a correctly made joint.

The data sheets will usually advise whether an alloy is suitable for a particular task under specific circumstances. For example, if items are soldered using a silver solder that contains too high a content of zinc, this is likely to lead to dezincification in a salt-water environment and the joint will become porous and weakened. In a case such as this, a silver solder with a lower percentage of zinc, or an alternative metal such as tin, should be chosen for the job. Another example is where stainless steel is

Silver Solders

	Ag	Cu	Ni	In	Zn	Sn	Temp. range (°C)	Tensile/ shear strength	EN No.
Silver-flo 56	56	22			17	5	618–652	410/165	AG 102
Silver-flo 40	40	30			28	2	650–710	450/155	AG 105
Silver-flo 20	20	40			39.9	0.1(Si)	776–815	330/145	AG 206
Silver-flo 5	5	55			39.9	0.1(Si)	870–890	390/135	AG 208
Argo-braze 56	56	27.25	2.25	14.5			600–711		AG 403
Argo-braze 632	63	28.5	2.5			6	691–802		AG 463

being silver soldered. If the resultant joint is in contact with water, it can lead to crevice or interfacial corrosion. Again, by the correct selection of silver soldering alloy this can be eliminated, or at least reduced to an acceptable level.

This table of silver solders shows the relative amounts of base metals in the alloys, the temperatures at which they are used and so on. The Argo-braze alloys are suitable for use on stainless steel as they do not contain zinc, which can lead to interfacial corrosion and joint failure.

SILVER SOLDERING OF BANDSAW BLADES

As proved time and time again, there is an exception to every rule. Usually it is not recommended to perform butt joints when silver soldering or brazing, but the following is that exception, or almost. To increase the area of the joint in the blade a scarf joint is ground on to each end of the blade. The bandsaw is a very useful form of saw, particularly for cutting wood, but also for cutting metals such as non-ferrous metals and the softer mild steels. The usual course of events when purchasing a replacement blade is to take the circumference measurement of the blade and buy a new blade at the correct length and right tooth profile for the intended cutting job.

The blade when manufactured is made as a continuous strip, with the teeth processed as the blade is made. A length of this blade equal to the circumference of that required is cut from the roll and is placed in a jig to hold the cut ends of the blade square to one and another, after making sure that both ends are true. These are now mostly resistance-welded, but were traditionally silver soldered.

Sometimes during use a bandsaw blade will break, usually at the joint, be it silver soldered or welded, or to one side of it where the heat from joining the blade together has changed the metal's structure. This can be

Bandsaw blade repair jig, made from bits out of the scrap bin.

due to the blade tension being too high, or simply that the wood or whatever is being cut is pushed through the saw too quickly, distorting the blade and causing it to snap. As mentioned earlier, the usual course of events is to purchase a new blade. However, if the blade is fairly new and is still retaining its sharpness and set, with the aid of a jig it is a relatively simple silver soldering operation to rejoin the blade, extending its useful life.

A simple clamping jig can easily be made from mild steel or whatever offcuts are in your scrap bin. There are no fixed criteria, other than the two ends of the blade have to remain parallel with the rest of the blade, so that it runs true and does not try to run off the bandsaw wheels. Also, these same ends must be lapped without any perceivable step, or the bandsaw may jam when cutting. To facilitate this reconnecting process, once the ends have been filed or ground square, both require a matching scarf to be ground on them. The best way to achieve this is with a belt sander. All that is required is to clamp the two ends in the jig before proceeding

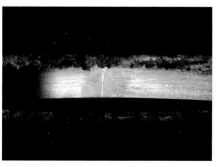

Silver soldered joint on a bandsaw blade.

with silver soldering. The cut ends of the blade will require fluxing where the silver solder is intended to run, that is, on the scarf joint itself. The two ends of the saw blade within the clamping jig will need heating to a cherry red, then the silver solder can either be applied to the joint, or a pre-cut length can be placed on the joint before heating. Once the joint has been filled by capillary action, the whole ensemble should be left to cool so that the clamp can be removed without burning yourself.

After the clamp has been removed, the joint can be cleaned back to a smooth finish with suitable abrasives, so that as the blade runs through the machine there is no step in the blade, or anything else to impede its travel. If the ends of the blade have been correctly clamped, the blade will run true. While cutting, the blade will be seen as a continual band without any wandering or juddering, provided that it is sharp and all the guides have been set correctly on the saw itself.

This perhaps is a good illustration of the strength of a silver soldered joint, whereby in this case the blade may be no more than 0.5–1mm in thickness, although with the scarfed joint this may effectively be as much as 2mm. Remember that it is flexing twice on every revolution around the two wheels of the bandsaw many times a second, plus where you are pushing materials through the saw, this puts more strain on the blade, albeit in another direction. The connecting joint therefore has to be tough to withstand this treatment.

SILVER SOLDERING PITFALLS

* Use the correct flux for the heating cycle.
* Ensure that the correct capillary gaps are present at soldering temperature.
* Avoid liquation of the alloy filler by rapid heating to melting temperature.
* Choosing the correct alloy for the job can be a problem.
* Not very good for butt joints.

A broken canopy tube on the garden swing chair required fixing and the only way was to fit a tube inside, as the break in the tube was across the screw hole that held it to the bracket. Any exterior sleeve would have hindered fitting it back together easily. Silver soldering was chosen to effect the repair, so as to avoid overheating the thin wall tube. It would require less heat then brazing, but would be more than strong enough for this job. Also shown is how to join a bandsaw blade, as described overleaf. Working in the right sequence will produce a strong join every time.

1. A piece of tube was turned down in the lathe to fit the inside of the broken tube. The paint and surface rust were cleaned off the parts.

2. A small quantity of Easy-flo flux was mixed with water and applied by brush to all the surfaces that were to be silver soldered.

3. After assembly and checking that the two broken swing chair arm parts were parallel to each other, the area was heated on a vermiculite block until the flux was seen to clear the surface to be soldered. The silver solder rod was then applied to the hot joint and the solder flowed into the joint.

4. Once the joint had been made, it was left for a while to cool. The soldered area was cleaned to remove any surplus flux and burnt paint.

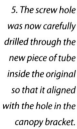

5. The screw hole was now carefully drilled through the new piece of tube inside the original so that it aligned with the hole in the canopy bracket.

6. After coating the exposed tube with primer and paint, the canopy was assembled and fitted back on to the swing chair, ready for further service.

7. Joining a broken bandsaw blade is not very difficult. The hardest part is keeping the two ends together whilst silver soldering. A clamping jig is required; this one was made up from bits out of the scrap bin.

8. To increase the area of the silver soldered joint, both ends of the blade were scarfed on a belt sander before aligning in the clamping jig, making sure that there was no step in the blade and that the two ends were square in the clamp.

9. Once the alignment of the two ends of the blade was satisfactory, the flux was applied on the joint and in this case a piece of cut silver solder was placed on the joint.

10. Gentle heat was applied with a blowtorch until the flux was seen to melt on to the blade surface, followed by the solder being drawn into the joint by capillary action. It is important not to apply more heat than is necessary to melt the solder.

11. The blade was left in the jig to cool down. This picture clearly shows that I placed too large a piece of solder on to the joint, as there is an excess on the blade surface. This all had to be ground off, wasting money as silver solder is expensive.

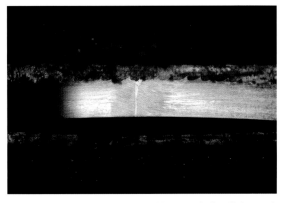

12. The finished joint, with the excess silver solder ground off and left smooth. The blade is now ready for further use in the bandsaw.

For efficient cleaning and preparation before any soldering or brazing task, a good supply of materials need to be at hand

6 Preparation, Cleaning and Pickling

PREPARATION

Where new materials are used for producing items to be soldered or brazed, cleaning before the operation should only require the minimum of effort to remove workshop soiling, general dirt, dust, fingerprints and so on, as well as the natural oxide layer that will have developed in storage. Abrasives such as aluminium oxide paper should be used to clean off the oxides, while any oil or grease can be removed with the aid of a proprietary degreasant, or even automotive brake and clutch cleaner. Of course, make sure that the lid is replaced on the can and that it is removed from the area where the heating is to be done.

Abrasives

There is a whole plethora of abrasives available today, from the traditional sandpaper through to the more modern aluminium oxides and silicone carbide papers. The advantage of the silicone carbide wet and dry abrasives, with their waterproof backing paper, is that they can be used wet, either with water or a solvent, which can help to keep the paper from clogging up with the removed materials. The finer grades, when used with water and soap, can be used to resurface paintwork before polishing with the fine polishes especially produced for this task. They are widely used in the car refinishing trade. A combination of plastics and abrasives has produced a new product, which is created by gluing abrasive particles to a mesh of plastic strands, not unlike a pot scourer. This is ideal for light rubbing duties and can be used both dry and also

A good selection of abrasive material, such as emery and silicone carbide papers, will help in cleaning back to bare metal.

Wire wool is available in several forms, such as preformed pads and on a roll, and in various grades from course to very fine.

with liquids, either water or a solvent. I have found the finest grade to be of particular use when spray painting between the last coats to remove any remaining nibs in the paint surface.

Wire Wool

Wire wool is a very useful cleaning aid that comes as a ready made pad, or on a roll. It is available in many grades. It is made from strands of steel and it is the edges of the steel strands that do the cleaning. A useful dodge is that the wire wool can be used as a carrier for a solvent such as cellulose thinners. By working the wire wool over the surface as you clean it, you can distribute the solvent right where you want it, speeding up the whole process. When soldering copper pipes and fittings with soft solder, sometimes all that is required before soldering is a quick rub with wire wool to remove any oxides.

When undertaking repair work, or where the components are made from recycled materials, extra precautions will need to be undertaken, as there may well be coatings of paint and or other protective layers that will need to be removed before anything else. Once these have been removed, oxide removal can be carried out. As already mentioned in Chapter 1, there are possible hazards associated with old paint coverings. Up to and during the 1960s paints more often than not contained lead, which is detrimental to health and the environment. Some of the more modern paints, particularly the ones giving the appearance of a hammered finish, use chlorinated solvents, and if these are heated they can give off dangerous fumes, not unlike mustard gas used in the trenches in World War I. So sensible precautions should be taken when removing these old paints, such as working outside and wearing a breathing mask.

CUTTING AND SHAPING TOOLS

Hand Tools

While this book is predominantly about the techniques of soldering and brazing, it would be incomplete without reference to the preparation of the materials to be used and the tools required to achieve this. The starting point is obviously the cutting out of the items to be soldered or brazed together. At the very basic level, a hacksaw will cut most metals, ferrous or non-ferrous, provided that they are not too hard. Hacksaw blades are available in various numbers of teeth per inch (tpi). The selection of the blade will be dependent upon the thickness of the material being cut, with coarse tooth pitch, 18tpi, being used for thicker sections, and the finest pitch, 32tpi, for cutting thinner sections or thin walled tubes and so on. At a minimum, at least two teeth need to be in contact at any one time during cutting, otherwise the blade will drop on to the material being cut between the teeth, with the usual result of broken teeth. One drawback with a hacksaw is that it will only cut in straight lines.

Sheet metal is most easily cut with snips or a guillotine, using a marked line as a guide; be careful of the sharp cut edges.

Once the object has been roughly cut to shape, it will then require fitting to the item to which it is to be soldered or brazed. The basic hand-powered option is the humble file. These are available in various shaped

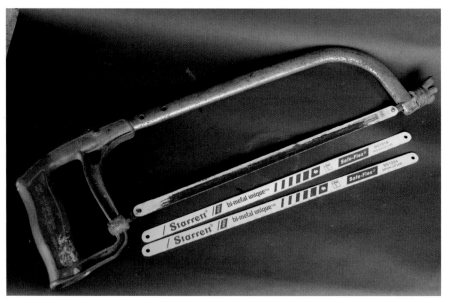

For hand-cutting metal, a hacksaw is very useful. A coarse blade should be used for thick materials and a fine blade for thinner or hard materials.

profiles, for example, round and half-round for forming curved shapes and of course differing grades, from the coarsest, single-cut bastard file through to extremely fine finishing files. Hand-sanding with aluminium oxide paper, silicone carbide sheets of wet and dry paper and of course emery cloth is possible. The wet and dry versions can be used with a lubricant, commonly water or paraffin/kerosene, which will help to stop the particles from clogging the grit too quickly and make it last longer too.

Measuring and Marking Out

Before any cutting out can be undertaken, the work needs to be marked out. This is an essential job to ensure accuracy, especially if the finished item is to hold pressure, such as a steam fitting. A good measure will therefore be essential. It can be metric or imperial, or both – it depends upon your age and preference. The main thing to remember is not to mix the two when working on a single marking out job, as although it is relatively easy to work out the conversions, swapping between the two will lead to inaccuracies. Normally I work in imperial measurements, but, for example, if I am working on meas-

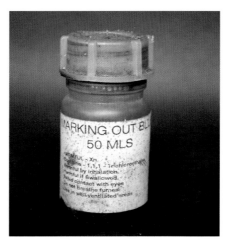

Marking-out fluid will be a help. Alcohol-based, it dries quickly and the marked lines will stand out against the blue colour.

urements that need dividing into several separate units, I find it far easier to divide them as a whole number, say 414mm rather than 16.299in. It is also worth remembering that when marking out several hole centres in a line, for example, that you should mark them individually from a reference point, such as the edge of the sheet, rather than marking the first centre and then measuring the second from this and so on. Although the error on each centre may be small, after

Once dry, lines and centre-punch marks will stand out on a surface coated with marking-out fluid.

several markings these small errors will be accumulative. By working from one reference there may still be errors, but they will not add together.

Other marking-out tools will include a square and a bevel gauge and of course if curves are required, a compass. To actually mark the metal, some form of scribe will be a good idea, as a fine scratched line is far more accurate than a fat marker-pen line. If a lot of detailed marking out is required and your eyesight may not be as it once was, a marking-out fluid can be used to help define the markings. This is a blue alcohol-based liquid that is applied using a brush or rag and quickly dries; once dry, the marking out is done on the blue surface. As the marking out is performed with a scribe, it scratches through the surface on to the metal below, showing through the blue. The contrast between the blue and the metal makes the lines more distinct and easier to see. Once all the cutting and any drilling has been done, the blue is easily removed with thinners on a rag so as not to contaminate the soldering operations.

The usual run of events when marking out is to scribe a faint line where it is required. The crossing of lines and the centres of any holes are marked with a light centre-punch mark. Do not make the centre-punch marks too deep at this stage. If they are slightly out of position it will be easier to remake the marks in the correct position, but if the marks are deep and large, remarking will be almost impossible. For long straight

When marking out multiple measurements, use one datum point to avoid an accumulation of errors.

lines, sometimes it is advantageous to mark along the length of the line with very light centre-punch marks at regular intervals, as sometimes the faint lines can be lost with subsequent work. With the centre-punch marks, the line position can easily be reinstated without further measurement. Once you are sure that the punch marks are in their correct positions, go over them again using a harder tap from the hammer.

The cutting lines marked are just that; when cutting, try to leave the actual line in place, as once gone it will be near impossible to know how much you have cut off without re-measuring. Once cut out with a saw, it will be easy to clean up to the line with a file or whatever, and of course more can be removed if needed upon a trial fit, but it is impossible to put it back on!

Centre-Punch Marks

The centre punch is useful for creating a capillary gap by raising an edge around the indent made by the centre punch on one of the two items being joined. Obviously the harder the punch is hit with the hammer, the deeper the indent and consequently the raised ridge will be bigger. Experimentation on some surplus material of the same type as is being used will be needed to get a feel for what you are doing, so as not to destroy your project.

Shears and Guillotines

Sheet metal can be cut with saws, but will be better cut with hand shears or a guillotine, up to perhaps a thickness of 1.5mm in non-ferrous metals such as brass, copper or aluminium, 1mm for mild steel and 0.8mm for stainless steel. Some will allow the cutting of curves with their design of curved cutting blades, but, if you are dexterous enough, it is possible to cut a curve with a straight-bladed shear, by turning the sheet at the same time as cutting. This can also work when using tip snips, especially the handed

varieties, known as aviation snips, which are designed to cut to the left or right, so that the cut material clears the hand on the cutters. A more accurate curved cut can be obtained by roughly cutting out to 3–5mm from the marked line and then re-cutting up to the line. This allows the waste to bend easily out of the way as you proceed along the line of the cut.

It is good workshop practice to keep a set of tools specifically for working non-ferrous metals, such as brass and copper. Once a hacksaw blade has been used to cut a piece of mild steel, for example, the blade will have lost the very keen edge that it had when new. It will still be fine for cutting further steel, but will make hard work of cutting the relatively softer non-ferrous materials. The same goes for files and so on. If the expense of a double set of tools cannot be justified, then at least keep a new blade or file handy for when any non-ferrous materials are to be cut or filed. Just remember to replace the blade with the old one before working with steel. Once the blade that has been kept for brass work and other non-ferrous metal use has lost its keen cutting edge, it can then be used for steel, a new one being purchased for brass and so on. The same applies for files.

Power Tools

Cutting and shaping can be hard work when performed by hand, especially if there is a lot of it to be done, but the benefit of using hand tools is that it gives you an insight into how the different materials respond to different actions and you can learn a lot about their individual properties. The obvious answer to this hard work is to use power tools, taking the effort out of the job and saving time, as most tasks can be performed by machine much more quickly than by hand and sometimes more accurately.

There is a whole range of power tools suitable for metal cutting and often at modest prices. Chop saws, the modern take

on circular saws, are now available with carbide-tipped blades that are suitable for most metal cutting tasks where the material is in strip form. As well as square 90-degree cuts, angles up to approximately 20 degrees can be just as easily set on the vice/clamp of the machine and, as the name suggests, chopped off. The now possibly outdated power hacksaw will also do an admirable job. This type of saw can be set up, loaded with materials and left to cut through at its own pace. Having a stop switch built in that automatically trips once it has cut through the metal in the machine vice, this type of saw will be an advantage if a lot of repetitive cutting is envisaged.

The main disadvantage with these first two types of saw is the fact that they will only make straight cuts. The answer to this particular problem is the metal cutting bandsaw, which will cut radiuses and circles and all forms of curves. Of course, it will just as easily cut straight lines, which means that it lends itself to producing irregularly shaped items to be soldered or brazed together. All that will be left to do is the final fitting to get the correct capillary gap with hand tools, or an angle grinder fitted with a sanding disc or flap wheel, taking off a little at a time and retrying the fit.

Oxy-Fuel and Plasma Cutting

If a lot of cutting of steel material is envisaged, then a quick and efficient way to cut it is with either an oxyacetylene or an oxypropane set-up, but this is rather expensive and of course entails a lot more safety awareness than with other cutting techniques. The easier approach is the plasma cutter. This equipment will slice through not only steel, but most non-ferrous metals as well. The process works by the use of an extremely hot plasma created by an arc between the end of the cutting torch body and a tungsten electrode, with the metal being cut, which is earthed. This creates a molten pool that is instantaneously blown away by a

When cutting with oxyacetylene or a plasma cutter, it can look very pretty, but those sparks are a real fire risk and can travel quite a distance across the workshop floor.

Here the angle grinder is fitted with a flap wheel sanding disc. These last much longer than a single sheet sanding disc. They are available in zirconium grit for aluminium and stainless steel cleaning, which avoids contamination of the brazing/soldering area with silicone carbide or aluminium oxide.

stream of compressed air, unlike the oxy-fuel process where the red hot steel combines with the pure oxygen stream, creating iron oxide before being blown from the kerf, or cutting line, by the high pressure oxygen.

The plasma cutter will produce a very neat cut edge and will also easily cut straight or curved lines and follow a template, so it is almost effortless to produce multiple copies of items to be soldered or brazed together. Again, safety is the watchword, as the hot particles blown from the cut will scatter all over the workshop floor, with potential fire risks from flammable materials lying around.

The Angle Grinder

The ubiquitous angle grinder can be placed into either the cutting or shaping power-tool heading. The angle grinder is used for cutting when fitted with a cutting disc and with care can cut radiuses, although it must be remembered that when using a cutting disc, which is relatively thin, no pressure should be applied to the disc perpendicular to its axis. The disc under these conditions will most probably crack and disintegrate,

with lumps of it flying all over the place. The other danger, if this happens, is that as a chunk drops away it will unbalance the rest, creating tangential forces that will try to rip the angle grinder from your hands. If alternatively fitted with a grinding disc or sanding disc, the angle grinder will happily shape metal to any profile. The biggest problem here when final fitting, will be the fact that it is easy to remove too much metal, so extra care is required when using it in this mode.

The modern approach to the sanding disc for fitting to an angle grinder is the flap

The ubiquitous angle grinder, shown here fitted with a thin cutting disc. It is especially useful for cutting aluminium and stainless steel, but can be noisy in operation, so wear ear protection.

wheel. This consists of many overlapping strips of abrasive stuck to a backing pad. The advantage over the ordinary sanding sheet is that it lasts a lot longer than ordinary abrasive sheets. For materials that are sensitive to contamination when soldering or brazing, such as aluminium or some stainless steels, these wheels are available made from a zirconium grit, which eliminates the

possibility of contamination from silicone carbide and other adverse contaminants. Perhaps not quite so obvious, but none-theless important, to prevent any cross-contamination do not use the flap wheel intended for aluminium and the like for steel.

As already stated, with a wire wheel fitted the angle grinder is an excellent tool for pre-cleaning and post-cleaning of items to be soldered or brazed. It is available in various grades of wire, giving a large selection to choose from to fit the severity and purpose of the job to be undertaken. As with all power tools, it is essential to wear suitable eye protection and a breathing mask, as the resultant particles will easily damage the eyes and lungs, being flung out from the machine at extremely high velocity. The individual wires from the wire wheels can also stick into the skin as they fly off at high speed, as they can penetrate right through clothing, From experience, this is very painful, so be careful. It would be advisable also to wear a full face shield for protection.

Bench Grinders and Sanders

For basic shaping of items, the humble bench grinder must not be overlooked. These power tools are cheap and readily available and of course have a multitude of uses around the workshop.

An alternative to metal shaping on a bench grinder is the sanding belt, which is especially useful on non-ferrous materials, and can be purchased fitted to one end of a bench grinder with a standard grinding wheel at the other, giving the best of both worlds. Various grades of grit are available for the belts for use on these machines, depending upon the task in hand. Dedi-cated finishing machines are available, but unless the workload justifies it, they will prove expensive.

The best of both worlds, one end a grinding wheel and the other a belt sander. This can be used for the shaping of parts to be joined. Various grit sizes are available.

Air Tools

For most electrically driven tools, there is an equivalent that can be run from an air compressor. Obviously the bigger the tool, the more air it requires to drive it. While looking at air tools, it is worth remember-ing that even a relatively small compressor is capable of running tools that require a lot of air, if only for a limited period.

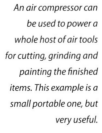

An air compressor can be used to power a whole host of air tools for cutting, grinding and painting the finished items. This example is a small portable one, but very useful.

Air tools cover a comprehensive range, with air-driven angle grinders, die grinders, air cut-off saws and so on. There are also percussion tools such as needle guns and air chisels that can be used for cleaning and cutting. Be mindful if using the latter, as they can be extremely noisy; although ear protection will reduce the noise for yourself, your neighbours will have little choice but to listen to it.

Once the soldering or brazing has been finished and everything has been cleaned up ready for a coat of paint, your compressor will come in useful again, with paint spraying equipment being available to run from it.

CLEANING POST-SOLDERING OR BRAZING

All soldering and brazing operations when completed, with the possible exception of electrical soldering, will require some form of cleaning to rid the formed joint and surrounding area of the flux used during the process.

Soft Soldering

With soft soldering, the remaining flux will be acidic, so there is the possibility of the flux continuing to corrode the items that have been soldered and anything that will be in contact with them. Cleaning of soft soldering fluxes is relatively simple, as most are in a paste form when used and will remain in a soft state after soldering. To remove these residues, all that is needed is to wash the items with warm water and give them a good scrub with a brush before drying, as the residue is water-soluble. Sometimes when a liquid flux, such as Baker's Fluid, has been used a black deposit will be left. This is only the dried flux with oxides and surface contamination that it has removed during soldering. It will not be very hard, although it will be acidic. It can also be removed easily by scrubbing with hot water.

Silver Soldering and Brazing

Due to the higher temperatures of silver soldering, fluxes that can withstand these temperatures need to be used. These in general leave behind a harder residue, although they are liquid when at brazing temperature. Soaking in warm water for 30 minutes or so and a good scrubbing will remove them. Once the higher temperatures of brazing are encountered, borax-based and similar fluxes will need to be used to be able to withstand the heat. The downside to these fluxes is that, once cooled, they are like a glass coating and are not water-soluble.

One procedure that will shift or at least crack this glass-like substance is, after brazing or silver soldering, to allow the alloy to solidify so that the joint cannot move. The whole ensemble is then quenched in a bucket of cold water. This has the effect of rapidly cooling the objects and the solidified flux. This rapid cooling shrinks everything and with the flux's glass-like nature,

Low-temperature fluxes can be removed by scrubbing with water; silver soldering fluxes may need to be soaked for 30 minutes first.

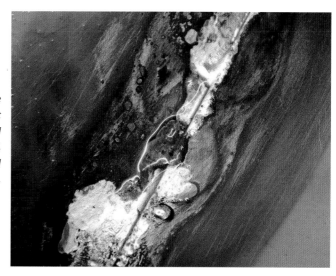

High-temperature brazing flux will set like glass and will need mechanical removal, such as by chipping and wire brushing.

it shatters into many pieces, making subsequent mechanical removal much easier. It must be remembered that this technique should not be used on anything like a pressure vessel, where the cold shock could cause weakening in the structure. Steels, other than mild steel, should also not be quenched, as this can harden them, making further work difficult, if not impossible.

PICKLING

As already discussed, when heating any metal, especially up to red heat, it will readily oxidize. This is why a flux is required during the soldering or brazing operation. Although mechanical means of this oxide removal will prove effective in many situations, sometimes where a decorative finish is to be left, scratching the surface with a wire brush or abrasive is not suitable or desired. In these cases, pickling in an acid bath will leave a chemically clean surface to be polished, or given further finishing coats of paint or other suitable finish.

Forty years or so ago every school metalwork classroom had a pickling bath, if my memory serves correctly, of 10 per cent nitric acid. This was corrosive enough to remove the oxidized scale from our copper tea caddy spoons and the like, but was not too corrosive for young teenagers to handle. Once having been in the pickle bath for the requisite time and rinsing under the cold water tap, any remaining scale could be removed with pumice powder. All that remained was to polish the said item on the polishing wheels. Contemporary Health & Safety would probably not approve!

Dry Pickling Salts

Although acids, such as hydrochloric acid and nitric acids, are readily available in industrial quantities, and will do a sterling job of removing oxides, these chemicals come with the inherent dangers associated with strong acid solutions and there-

The use of straight acids for pickling is no longer recommended. Today, pickling salts are readily available that are pretty much inert while in dry crystal form. Mixed with water for use, they will lift oxide scale and help to soften flux residue, as well as removing corrosion.

fore cannot be recommended for the home workshop. The modern approach is to use pickling salts. These are much more user friendly and safer in the home workshop environment. The pickling salts come as dry crystals, so are easy to store and are fairly inert in this condition. The active ingredient is usually sodium bisulphate, which, when mixed with water, is mildly acid and will readily clean items placed in it. The process I believe can be speeded up somewhat by warming the resultant solution.

One of the advantages of dry storage is that only the required amount needs to be mixed for the job in hand, leaving the rest for another day and another job, plus the chemical is more or less inert until activated with water, although the crystals should not be stored in any ferrous metal container. When mixing the crystals with water it will be best to use a plastic container. Although the acid should not affect glass, if glass is used there will be a danger of accidental breakage and the acid spilling all over the workshop floor or bench.

Mix and use pickling salts in a plastic container, reading the supplied data sheet to determine the solution strength. Using the solution warm will speed up its operation.

Although pickling salts are less corrosive than pure acids, precautionary preventative measures should be taken when handling them, whether wet or dry. Appropriate pvc or nitrile gloves should be worn, and when placing items into the pickling bath any splashes on to the skin should be washed off immediately with soap and water, as they will possibly ulcerate and certainly irritate if left. Eye protection should also be worn, as the eyes are vulnerable to splashes, with any salt entering the eye causing conjunctivitis. Care should be taken to dispose of the used pickling salt bath correctly: read all safety information concerning handling and disposal supplied with the product.

PUMICE POWDER AND ABRASIVES

Pumice powder is used after the pickling bath and the items have been rinsed thoroughly. This is basically just a fine abrasive powder, used to clean off the loosened oxide scale, leaving a clean finish ready for painting or whatever finish you require. Pumice powder is a natural product, derived from rocks formed during volcanic eruptions. The pumice is broken down into smaller and smaller pieces until it is a coarse powder, which is then further ground into varying grades. It has many commercial applications.

For small jobs, pumice powder can be applied with a dampened rag, but for larger jobs the pumice can be mixed with water to form a slurry, which is then applied with a suitable rag. Once the scale has been removed, a good rinse in water will reveal the clean metal ready for the next process. Copper and brass, or indeed most metals, can be polished to a nice finish, once the scale has been removed, and can look extremely decorative.

To remove any remaining remnants of scale left from heating, or just to clean, pumice powder is used mixed with water. This is effectively a fine abrasive made from volcanic rock, available in several grades.

The pumice powder is mixed with a little water into a wet paste. Use a cloth to apply the slurry and work in all directions until all traces of scale have gone.

Once the scale has gone, leaving a matt finish, rinse off in clean water and dry the item, especially if made from steel as it will now readily rust.

POLISHING WHEELS

Hand polishing is possible, but unless you have a huge amount of spare time it makes sense to invest in some form of mechanical polishing. At the bottom end of the spectrum, polishing wheels are available to fit any electric or battery drill, but if serious polishing is envisaged a more appropriate approach will be to invest in a polishing machine. These are not unlike an off-hand bench grinding machine, but instead of having two grinding wheels and their necessary guards, one at each end, they have instead two polishing mops. One will be for a coarse polish (cutting) and the other for a finer polish (colouring). The usual way that the polishing mops are attached to the machine is that each end of the shaft has what is known as a 'pigtail' (a tapering thread), on to which the mop is mounted, one end having a right-handed thread and the other a left-handed thread, so that as

the machine revolves it tightens the mop on to the thread. Once the machine is stationary, it is easy to unscrew each polishing mop from its respective pigtail, making mop changing easy.

The polishing compounds are available as a solid bar, not unlike a bar of soap, in various grades and for differing metals. These are held against the polishing wheel, so that the heat generated by the friction of the bar of polish against the cloth mop melts the compound. The polishing compound then sticks to the mop. The skill is in getting the right amount of polish on to the wheel, as too much can be as much of a hindrance to the polishing effect as too little. As the item to be polished is lightly pressed against the now loaded polishing wheel, the abrasive within the polishing compound will cut away at the surface. Once the coarse polish has been used, the second wheel can then be loaded with a finer polish and once all traces of the coarse polish have been removed from the object being polished,

Decorative items can be enhanced by polishing. Mops, wheels and polishing compounds are available for use on different metals and for varying levels of finish.

A high degree of shine is possible if the grades of cutting compounds and polishes are used in the correct sequence, each one finer than the previous.

metals, through to stitched cotton mops in various grades for coarse buffing, and on to unstitched calico mops for colouring work.

The polishing itself involves just scratching the surface of the object being polished. Once all the scratch marks of the coarser grade have been eliminated by the finer scratches of the polish being used, you can move on to the next finer grade of polish. Although it is probably obvious, due to the high-speed revolving mops and abrasive compounds that are being used, make sure that you wear safety glasses, goggles or a face shield while undertaking mechanical polishing, as a speck of polishing compound in the eye will have a detrimental effect on your eyesight.

the process can then be repeated on this second wheel until the desired shine is reached. More polishing compound may need to be applied as the process continues.

It is important that the different grades of polishing compounds are kept to their respective polishing wheels, as cross-contamination will not help the polishing procedure. If the fine wheel is inadvert-ently contaminated with the coarse polishing compound it will ruin the nice shine brought up by the fine polish. This will entail thorough cleaning of the polishing wheel, or purchasing a new one and keeping the contaminated one for coarse work. Polishing mops are available in many grades, from mops made from sisal, used for fast material removal especially on ferrous

---

GENERAL PITFALLS

* Power tools can be noisy.
* Power tools increase the danger of accidents.
* Dangerous fumes from lead-based paints.
* Fire risk from sparks associated with oxyacetylene and from plasma cutting.
* Acid burns from pickling baths.
* Hand tools can be slow.

Before any soldering or brazing is to take place, materials need to be clean and free from oxides, such as rust and old paint. This is best removed before cutting out pieces to be joined, as it is easier to hold a larger piece. Once the joint has been made, cleaning up is required before paint and so on is applied.

1. Cleaning for any soldering or brazing operation requires that the objects to be joined are thoroughly cleaned. A wire wheel on an angle grinder will shift most of the rust and old paint.

2. Once most of the surface contaminants have been removed, a flap wheel will remove any remaining corrosion, leaving shiny metal ready for soldering or brazing.

3. The angle grinder fitted with a cutting disc makes short work of cutting out materials. The downside is it is noisy and produces a lot of sparks.

4. Hand tools such as this second cut file have a place amongst the power tools. The file was used to level out the old pipe boss before silver soldering on a new one.

5. With smaller components, the use of power tools could prove difficult. Here, a wire brush was used on a small component to remove the oxides from a previous heating operation.

6. Silicone carbide paper was to clean up the new fuel pipe boss. Trying to hold this to use a power tool would be near on impossible.

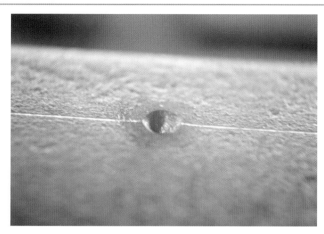

7. It is important for accurate work that marking out is done in a meticulous manner, leaving little room for error.

8. The old adage 'measure twice, cut once' is perfectly true.

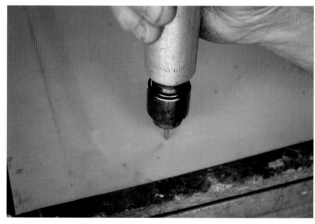

9. Air tools have a useful place. Here, a small air drill was used with a centre drill to drill the corners before the folding of the brazing hearth tray.

10. The belt sander is an invaluable tool for accurate sizing and removing the inevitable burrs on the edges of sheet metal.

11. The wire wheel has a vital role in any workshop. It is ideal for initial cleaning and removal of stubborn flux remnants.

12. Once all the soldering and brazing is finished, a good shine can be created on brass and copper with the use of polishing mops and fine abrasives.

7 Soft Soldering

Although not used as extensively now as in the past, soft soldering has still got a useful place in the available workshop techniques at our disposal. Its loss of popularity has most likely come about because of the change of materials in things that were once made from metals, such as tin plate and brass being superseded with plastics and, of course, with the fact that items containing lead can be a health hazard. As a result, this has led to a phasing out of the techniques used for joining everyday objects.

Soft soldering is a fairly simple technique to master. Most soft soldering is carried out using some form of soldering iron, electric or heated, from an external heat source such as a blowlamp. The usual soldering iron consists of a copper bit, with the size very much depending upon the size of the job being undertaken. Obviously a large bit will take more time to heat up, but that heat will last longer than with a smaller bit.

A large electric soldering iron – very useful for small to medium soft soldering jobs.

A large copper iron, which can be heated by a gas blowlamp, making it portable. This is useful for repairs away from mains electricity.

OPPOSITE PAGE:

Soft soldering is a dying art; the only real application left today is in plumbing.

The only real day to day soft soldering these days is on plumbing fittings and even these are being superseded by plastic. Here is a fluxed pipe joint ready for heating.

As the flux boils, add the solder into the joint until a completed silver ring is seen around the fitting.

The finished joint. All that is required is to clean the remaining flux to give a joint that should last indefinitely.

As with all types of joining of metals, be it welding, brazing or soldering, the effects of heating the metal to form the joint causes oxidation of the metal surface. During welding this manifests itself by producing brittle and or porous welds; with soldering it stops the solder from adhering to the metal surface. From this, it follows that in soldering in all its forms cleanliness is a must, with all surface contamination on or near to the soldering site being cleaned off, along with any surface oxides that may already be present.

Once the metal to be soldered has been cleaned to a bright, shiny condition, the flux needs to be added to prevent the work carried out in cleaning being wasted. The usual next step is to tin the areas to be joined, or for ferrous metals that can be more difficult to tin, a tinning paste is available, which contains flux and fine solder particles in suspension. The solder used for the particles is usually of a higher tin to lead ratio than ordinary solders, or for lead-free soldering it is available made from tin only. This is where the term 'tinning' originally came from, as tin has a higher affinity for ferrous metals and produces an easier bond. Once tinned, the solder will readily bond with the surface. With copper and its alloys, such as brass and so on, the tinning process can be dispensed with and the fluxed surface can be directly soldered. For example, when soldering copper pipe fittings in domestic heating systems, all that is required is to clean the pipe and fittings, then flux and assemble, ready for soldering.

Once the joint to be made is heated, the flux will appear to boil from the joint; adding solder at this point, it will be seen to be drawn into the overlapping joint by the act of capillary action, with a silver ring around the joint interface showing that the joint is complete and satisfactory.

Historically, tin-plated mild steel, known as tinplate, was used extensively in liquid-containing vessels. All manner of fuel tanks, cans and the like were manufactured in huge quantities; toys were also made from it. The

A soldered brass car radiator; even these are becoming a rarity as the end caps are mostly now made in plastic.

The essential thing when using a soldering iron, be it large or small, is the rapid transfer of heat to the joint to be made. A larger soldering iron will retain more heat and have more available to be able to solder for longer, which is an advantage in sheet work where it is preferable to complete the seam in one go. That said, reheating of the iron and reapplying to the part-soldered joint should not be detrimental if done carefully and time is allowed for the reheating of the end of the joint already soldered, so that the solder is in one continuous joint.

Soft soldering can be performed satisfactorily with a naked flame if carefully applied in certain circumstances, but the right flux will need to be chosen, as some fluxes do not like the high direct heat of the flame and will consequently break down and be ineffective. This will leave a poorly made joint, which of course will entail starting all over again after removing the burnt flux.

If working with new or clean materials, it may be worth trying a normal-grade paste flux, one that is sold for use with plumbing fittings. This should be aggres-

advantages were that the tinplate gave a degree of corrosion resistance to the underlying steel and the tin helped in the soldering of the joints. The canning industry grew into large companies, with cans being made from tinplate and sealed along the joints by soldering. With the contents being sterilized by heat before the final seal was made, the food kept fresh for many months. However, once again with the discovery that lead was poisonous, soldered tins, especially with solders containing lead, were abandoned. Nowadays, a lot of tin cans have a plastic coating on the inside, so as to avoid contact between the tin and its contents, plus many food containers are made from plastic.

Zinc sheets were extensively used for lining water tanks and could be used instead of the traditional lead flashings in the building industry. Zinc is easy to solder to make watertight joints, as it has an affinity to lead and the solders made from it, and the other advantage of using zinc instead of lead is that it is lighter.

Another traditional use for soft soldering was the manufacturing of automotive radiators, made from brass and copper soft soldered along the seams, but, once again, this has been largely superseded by the use of modern plastics for many of the components. However, the critical heat-dissipating qualities of brass and copper are still used for the radiator core, although lightweight aluminium is also often used these days, bonded together by modern glues and resins.

Soldering using Baker's Fluid is the same as using any other flux, although it is more aggressive at cleaning as it is very acidic. Be careful where you put it.

It is vital when you have used an acidic flux to wash the flux off thoroughly after you have made the joint; leaving it will allow the acid to carry on corroding.

sive enough to lift very light oxidation after a good rubbing with wire wool or other suitable abrasive. If repair work is being undertaken on dirty and old items, such as trying to repair a leak in an old car radiator, a more aggressive flux will need to be used. The most widely available is the proprietary brand called Baker's Fluid, which is made from hydrochloric acid into which zinc filings have been dissolved to saturation point. This has the effect of moderating the action of the acid, but leaving enough for thorough cleaning of the joint as it is made. Baker's Fluid is applied with an old brush along the line of the joint just before it is made. Keep the brush for this purpose only, as once placed in the fluid it will be very acidic. Moving the soldering iron along the joint will create a lot of hissing, as the fluid vaporizes from the heat of the soldering iron. Once the joint has been satisfactory made, both sides of the soldered joint and anywhere the fluid has run during the operation will need to be scrubbed with water to remove the acid.

PRODUCING A LAPPED JOINT IN SHEET METAL

The basics have already been discussed to make a satisfactory joint; putting these together to make a lapped joint will now be explained. Whatever form of soldering iron is being used, be it an electrically heated soldering iron or a gas-heated copper bit, the technique is the same for the actual soldering. After cleaning and tinning the respective edges to be joined on both panels, the two need to be aligned into their finished positions and held so that the tinned surfaces are together. Flux is then applied in and along the joint. This is easiest done in the horizontal plane on a workbench with assistance from gravity.

While everything is being sorted, make sure that the soldering iron is heating up. Once it has reached the required temperature, which will be evident upon tinning the iron, the heated soldering iron tip is placed against the top sheet, wait a few seconds to allow the heat to transfer from the iron into the sheets. The flux will at this stage be sizzling; this is when to apply the solder to the joint. Next, the iron is slowly moved along the line of the joint, heating the joint as more solder is flowed into it. On reaching the end of the joint, the soldering iron is removed from the work and the solder flows into the joint. Once cooled, the sound joint can now be cleaned of flux.

When soldering a lapped joint, the edges will require tinning before being placed together to solder. Here, tinning paste is used.

CONTAINER REPAIR

Containers that are made from a solderable material, such as sheet metal, brass or mild steel, can easily be repaired with soft solder. It can also be used to repair areas where a rust hole has appeared, possibly due to water from condensation if it is a fuel can. If the hole is just a pinhole, this can be cleaned thoroughly and the hole filled with solder. However, a better and more substantial repair can be carried out, especially if the hole is somewhat larger, by cleaning the affected area and then tinning. A suitably sized patch can now be prepared and the side to be in contact with the container also needs to be tinned. All that is now required is to solder the patch on, placing the soldering iron in the centre of the patch until the fluxed patch starts to sizzle. Solder is then applied to the edge of the patch and will run around the whole periphery of the patch. Once the flux has been cleaned, a coat of protective paint can be applied for longevity.

One note of caution is that if the container has contained a volatile liquid, such as a fuel of any kind, use a soldering iron by all means, but keep the container away from the gas torch as the fumes within the container will make an effective bomb if the fumes ignite.

Once the edges have been tinned, they are placed together using clamps.

A soldering iron is applied along the joint while feeding in solder. As the metal heats up in front of the iron, the solder will be seen to be drawn in.

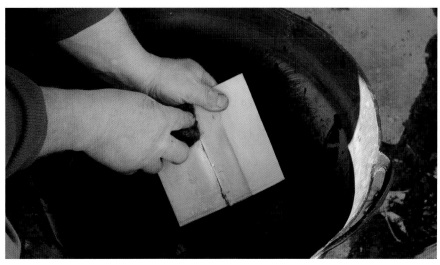

After the soldering has been done, the flux residue will need cleaning from the finished joint.

A common problem is a hole in the base of an otherwise good metal can.

The simple solution is to solder a patch over the hole.

SOLDERING COPPER PLUMBING FITTINGS

There are various ways to solder the copper plumbing fittings that are used in domestic central heating and hot water systems. The fittings come in a variety of forms, including the compression type that only require tightening with a pair of suitable spanners, but soldering fittings are available in two types. One is the 'Yorkshire' fitting, where a ring of solder is built into the fitting and is sufficient to complete the joint, and the other is what is known as end-feed fittings. These are plain fittings that are a snug fit over the pipe and the soldered joint is made by feeding in the solder from a reel from both ends of the fitting. With these, operator skill is required to judge how much solder to apply to each fitting for an effective joint to be made.

Whether Yorkshire or end-feed fittings, they both require heating up to soldering temperature. The traditional method is to use a blowlamp, but a specialist type of electrical soldering iron is available specifically for this job. To look at, the soldering iron is not unlike a pair of large pliers, with the jaws shaped to fit tightly around the pipe fittings. This keeps the heat generated from the soldering iron right where it is needed and will eliminate the risk of fire, especially if work is being undertaken under floorboards where a plethora of inflammable detritus exists, usually left by the builder.

When using the pre-soldered Yorkshire-type fittings, after cleaning the fitting and the ends to be placed into the fitting (usually a rub with wire wool will suffice), a small amount of suitable flux is applied to the fitting and then placed on the ends of the pipe. On heating, watch for a silver line appearing at the end of the fitting. Once all the way around the pipe, you can be certain that the solder has filled the capillary gap between the pipe and the fitting, making a sound, watertight joint. After removing the heat, the pipe needs to be left to cool

Yorkshire-type plumbing fittings are distinctive by their raised bands at each end containing the solder to make the joint. It can just be seen in several fittings.

before moving, otherwise the joint may move before the solidus point of the solder has been reached, which might the joint unsound.

When using end-feed fittings the technique is the same, except that once the joint has heated up to temperature the solder from a reel is fed into it. You will see the solder run around the whole of the joint, indicating that it will be sound. It probably does not matter whether the flux is applied to the fitting or the pipe, but if the flux is applied to the fitting, as the pipe enters the fitting the flux will only go on to the pipe where it is needed. If the flux is applied to the pipe end, as the pipe enters the fitting the chances are that the flux will be scraped off the pipe by the fitting. Although enough will be left to make a satisfactory joint, the excess will, on heating, run along the pipe and could lead to the solder, as it is applied, following it. This will make for an untidy looking job, although it will still probably be perfectly watertight.

At this juncture, it may pay to explain that too much flux will give possible joint problems just as much as too little flux. An excess of flux within a joint could impede the flow of the solder, preventing the capillary action from completing the ring of solder all around the joint, especially with the modern lead-free solders, as they do not flow quite so readily as the traditional plumbing solders containing lead.

Another point worth mentioning also, concerns the use of wire wool to clean the pipe and fittings. It is possible to trap a single strand of the wire wool between the pipe and the fitting. Once this has been soldered all will appear sound, but I have seen the result several years later where this strand of wire wool, being made of steel, has rusted through, leaving a very small passage through the soldered joint and thus causing a leak. The tell-tale sign that this was the cause was a rusty mark from the point where the water was dripping.

End-feed fittings need solder fed in from the end during the soldering process.

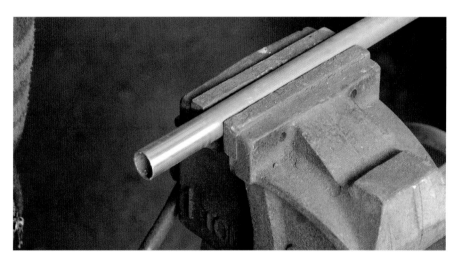

A Yorkshire fitting should be cleaned with wire wool before placing together.

The fluxed is joint ready to heat; in practice, several fittings can be prepared before lighting the blowlamp.

When heating the joint with the blowlamp, the flame is moved all around the joint, spreading the heat evenly.

Unsoldering a Joint

Sometimes a soldered joint will need to be undone and remade. Perhaps the joint was dirty and did not seal completely, or perhaps the angle of the fitting was wrong, or in the wrong place. It is always worth cleaning around the joint to be undone, even if it has just been completed, as an oxide layer may have built up from the previous operation and there will possibly still be some residual flux. In essence, treat it as if you were making a completely new joint.

Before heating, it will be a good idea to flux the joint, as this will avoid the solder already in the joint oxidizing excessively as everything heats up. It will be found that more heat will need to be put into the joint, as some of the alloy in the solder will have formed another alloy with the copper in the pipe and fitting. Once the solder is seen to be in a liquid state, with a twisting, pulling action, the fitting can be removed from the pipe.

Once cooled down, if the fitting is to be reused it will require cleaning. There also might be some solder adhering to the fitting that will need to be removed, as this would impede putting the fitting back together. After cleaning and possibly scraping, the joint can be remade, treating it as if it is a new one, but making sure that this time it is at the correct angle and so on before heating.

Here, the shiny solder ring around a finished joint can be seen. If this is complete, it is almost certain that the joint will be secure.

SOLDERING ALUMINIUM

It will be pointless soldering aluminium with ordinary solders containing zinc or lead. Even if the joint is successful, the chances of it lasting are very remote, especially if any dampness is around the joint. This moisture, in particular if there are any salts around, will react with the zinc or lead at the interface with the aluminium by electrolytic corrosion and within a matter of days the joint will fail completely. The biggest problem with aluminium when trying to join it or stick anything to it, is the tough oxide layer that forms on its surface. Conversely, it is this layer that gives aluminium and its alloys their corrosion resistance, as once the layer has formed it helps to prevent further corrosion.

Proprietary solders are available with a flux core specifically for soldering aluminium and its alloys. This patented range of solders called Alu-sol will successfully adhere to most aluminium alloys and also some types of stainless steels, although they will not adhere to anodized finishes or aluminium containing silicone. They can be used with a soldering iron or gas torch, although care will need to be taken if heating with a high-temperature torch such as oxyacetylene, as aluminium melts before any warning change of colour occurs.

Sometimes it is necessary to unsolder a fitting, especially if it is in the wrong place or at the wrong angle.

Once the parts have been cleaned, they are ready for resoldering.

SOFT SOLDERING PITFALLS

* Older soldered joints may contain lead.
* Joints produced are not as strong as silver soldered ones.
* No good if exposed to high temperatures or excessive vibration.

I am demonstrating the repair of a small auxiliary fuel tank from a vintage tractor using soft solder. Under normal circumstances, a flame should not be used near a fuel tank. However, in this case the tank had not contained any fuel for years and I washed it out with detergent before proceeding. The second half of this practical feature shows a vintage radiator being repaired. A crack had formed in the soldered joint through vibration over many years of work.

1. This fuel tank had already been repaired at some time in the past, but either it was not a successful repair or further corrosion had taken place, as it leaked.

2. After the old solder was cleaned off, the hole in the tank could clearly be seen. Thorough cleaning back to shiny metal was the first step in preparation for the new solder, getting rid of any corrosion.

3. Materials required for the repair; tinning butter would be used to tin the surround before filling the hole with soft solder.

4. Tinning butter was painted on to the area to be tinned with a brush. It has a matt grey appearance.

5. A flame was gently played over the area until the tinning butter took on a silver appearance. Once wiped off, a shiny tinned surface would be revealed.

6. The repair area was fluxed and reheated and solder applied from a stick or reel to cover the hole.

7. This radiator top hose fitting had a crack in the soldered joint, which needed to be remedied before replacing on the tractor.

8. The brass radiator would draw the heat fast, so I used the biggest soldering iron I had, which takes a few minutes to heat up with the blowlamp.

9. Once the iron was hot, it was plunged into the flux for a second.

10. On withdrawal from the flux, the iron was tinned with solder.

11. I had cleaned around the damaged solder joint beforehand; after holding the iron on the area for a second or two, the solder began to flow.

12. The finished repair. Hopefully this will hold for another fifty years once it is back on the vintage tractor.

8   Lead Loading, or Body Solder

The easiest way to get started with lead loading is to purchase a kit. This will contain all that you need except for the heat source and the body file.

During the early years of automobile ownership, possibly up to the 1960s, if your car, van or whatever form of transport you owned had the inevitable 'ding' or mishap, the only remedy to effect a repair was to bash out the dent, weld in a patch or new panel, then use lead loading to smooth out the panel work and bring it back to the original shape, or to hide the seams of the new panel. Sometimes large or awkwardly shaped panels were made in several pieces and then welded together and some of them were available as repair panels. This all changed with the advent of plastic body fillers; their ease of

The heat source used in this chapter was the standard propane blowlamp, although alternatives are available such as an acetylene and air torch, which gives a very soft flame.

OPPOSITE PAGE:

Although lead loading may seem like a bit of a rigmarole when compared to more modern techniques, it will be long lasting and on vintage and classic vehicles is the authentic way to proceed.

use meant that virtually anyone could fill a dent and sand it back with very little skill required. This was okay up to a point, but it also became the favourite bodge of the back-street garage, hiding a multitude of dodgy work.

BEWARE THE BODGERS

The 'technique' was to fill any large gaps or holes with crumpled newspaper or anything else to hand before trowelling on the filler. The same could be done with rusty and rotten panels, with the filler initially sticking to them readily. This would then be passed off as high-grade repair work with nothing outwardly showing to the contrary. The effects of the weather, along with the general wear and tear on the shabby repair, plus the problem that some fillers will absorb water if left uncovered, would eventually reveal the poor quality of the workmanship, unfortunately long after the vehicle had left the garage. This has led to the modern filler's unenviable reputation in

The bodger's 'stock in trade' – all manner of things can be hidden behind a skim of filler.

some quarters, of only being good enough to carry out bodged repairs with little hope of them lasting. Although beyond the remit of this book, polyester resins and fillers when used correctly and with the right preparation can lead to satisfactory and more or

less permanent repairs being made and of course without the inherent difficulties of lead loading.

It is virtually impossible to do a bodged repair with lead loading. The surface on which the repair is undertaken must be

This is the result of a 'bodged' repair with polyester filler; moisture from under the vehicle wing has caused the filler to bubble up through the paint.

Now a classic car, the Austin A35 used lead loading to hide panel joins in its construction; these are still in good condition some fifty years later.

Lead Loading, or Body Solder ● 103

The golden rule when lead loading is to make sure you have thoroughly cleaned back to bare metal before moving on to the next stage.

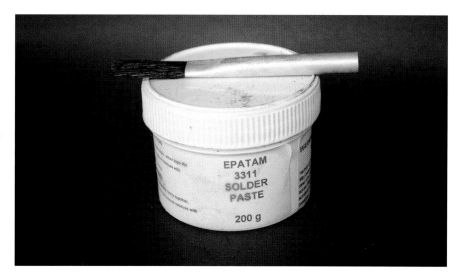

The tinning solder, a mixture of fine solder particles and flux, is painted on to the panel with a brush.

An even layer is applied to the panel. This will give a matt grey appearance.

The matt grey surface is heated gently with the blowlamp flame until the surface has turned a distinctive silver as the solder melts. Wiping while still hot should reveal a continuous colour on the panel surface.

The surface of the tub of tallow is heated with the blowlamp. Dipping the wooden paddle into the melted tallow will stop the solder from sticking as it is pushed into shape with the paddle.

clean and free from impurities, so as to enable the lead loading alloy to bond with the metal. Failure of this bond would mean that the solder will not 'stick' to the surface, so it follows that a lead-loaded repair cannot be compromised, as with plastic fillers. It will have to be done properly, or not at all.

Many early cars had some of their seams covered by lead loading at the factory. The Austin A30 and A35, for example, had the seam where the roof joined the body at the rear lead loaded after being welded, the lead alloy just smoothing the transition from one panel to another. It had no intrinsic strength of its own, acting as just a filler, but had the advantage that once the seam was filled, it was watertight. Modern vehicle construction techniques have moved on, with very few seams requiring hiding from view, as there are very few used in the building of a modern car; this is probably down to superior presses that can press out larger panels in one go, rather than building up a large panel from several smaller ones.

The actual technique is the same as any soft soldering. To get the solder to adhere to the surface to be repaired it must first be adequately cleaned so that the subsequent tinning of the surface is complete. Once tinned, the difficulties start; the biggest problem is that you are working with gravity trying to pull the body solder off the panel before it has solidified. If it solidifies too soon, it will require reheating to get it moving again. Too much heat and the whole lot will dribble on to the workshop floor and the process will have to be started all over again. This is where skill and patience come to the fore, learning to apply just the right amount of heat to soften the solder without it becoming runny.

In previous chapters, the liquidus and solidus points of the soldering alloy have been discussed and with certain processes it is advantageous to have a eutectic alloy where the liquidus and solidus tempera-tures are one. In the case of body solder it is advantageous to the operator to have a wide differential between these two points, as it is during this portion of solidity of the alloy and the liquefying point that the oper-ator can work the solder like butter. It can be manipulated with wooden paddles into the desired shape, hopefully without it drop-ping off the panel on to the floor.

It is, of course, not only the body solder that is hazardous to the human body. Due to the cleaning nature of fluxes, they are acidic. But provided that suitable precautions are taken, the dangers are minimal and the flux must be thoroughly washed off the panel work before any further work is undertaken, such as shaping and painting.

APPLICATION TECHNIQUE

As stated above, it is almost impossible to do a bad job with lead loading – that is, while you might find the task too difficult to com-plete, if you are able to complete everything

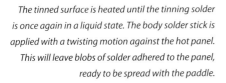

The tinned surface is heated until the tinning solder is once again in a liquid state. The body solder stick is applied with a twisting motion against the hot panel. This will leave blobs of solder adhered to the panel, ready to be spread with the paddle.

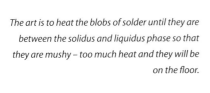

The art is to heat the blobs of solder until they are between the solidus and liquidus phase so that they are mushy – too much heat and they will be on the floor.

as it should be done, then the result will be a good finish.

If it is crash damage that is being repaired, it will be no good trying to slap the lead over as is. The majority of the dents will need to be dressed out beforehand, leaving the lead loading to take up any minor irregularities in the surface and the lead shaped to the final contours with files. For obvious reasons, power tools should not be used on lead loading as the high-speed abrasive action will send a cloud of fine lead particles into the air, which will be detrimental to you and any bystanders in the vicinity.

To make a successful job of lead loading, it is imperative that the base metal is scrupulously clean before starting the work. The first operation is to tin the surface so that the subsequent solder will adhere without any problems. This has been made easy with modern advances; a tinning paint has been developed that contains solder with a high tin content suspended in a flux medium, so that once painted on to the surface to be body soldered, it will more or less stay where it is put, ready for heating.

Once the tinning paint has been applied evenly, the whole area is gently heated with a blowlamp of a suitable size for the job in hand. As the panel comes up to temperature, the particles will be seen to melt on to the surface and fuse together. An old but clean rag is then used to clean over the area being worked on. This will wipe off any excess flux and solder and should also reveal an evenly tinned panel ready for the next stage. If, however, the resultant tinning looks patchy, where it has not adhered to the panel, it will have to be redone, or at least the parts where the tinning has failed to take.

The body solder comes as a stick, 25 × 6 × 300mm, and possibly the hardest job is to get the solder from this stick on to the panel. The correct technique is to heat the panel gently until the tinning can be seen to glisten as it begins to melt. The solder is applied by pushing the stick of solder square on to the panel, giving it a twisting motion at the same time. If the panel is at the correct temperature, a dollop of solder will be left adhered to the tinned surface. This needs to be done all over the area being worked on and a careful judgment has to be made as to when there is enough solder on the panel to bring it up to the desired contour.

In body soldering, the solidus and liquidus points of soldering alloys are exploited to our advantage. While the body solder is kept between the solidus and liquidus points, it will be in a mushy state, almost like a soft plastic. This allows us to manipulate the solder on the panel to the contours required. Wooden paddles, often made from maple, are used to push the solder to where

With the paddle, the blobs are manoeuvred to the desired contour, with heat applied as necessary to maintain the mushy state.

Once the panel has cooled, it needs to be washed off water and dried. The solder can then be shaped with a body file, with the final contouring done by hand, as mechanical means will release fine lead particles into the air.

With the lead firmly bonded to the panel, you can be certain that it will not bubble up as with polyester fillers and will take any paint finish that is desired.

If inadequate cleaning has been performed, areas in the silver tinning will show through as black patches. The only remedy is to start again, as the body solder will not adhere to the untinned areas.

mushy and can be pushed around into the required shape. It may be best to go over all of the solder dollops roughly, gently heating each in turn, so as not to put too much heat into the panel. Withdrawing the heat in-between will allow some cooling to take place. Once a rough contour has been achieved, careful heating of particular areas can then be carried out, refining the contours. When happy with the result, the panel should be allowed to cool completely, any remaining flux washed off and the panel dried.

To finish the job ready for painting, the resultant lead loading will now need to be shaped with a body file. These are available in various grades, but for the soft, or relatively soft. solder, a coarse-cut blade will suffice. The blade is held in an adjustable frame with a handle; this allows the tension on the blade to be adjusted so that the curvature of the blade matches that of the panel. As the solder is soft, the body file pares the solder off, rather like cutting with a knife. The result is curls of solder rather than dust and a smooth finish.

Once the contour is as near as you can get it and you are satisfied with it, it can be hand-sanded to prepare it for painting, safe in the knowledge that once the paint is on, the underlying solder will never lift and bubble up, as can happen with modern polyester fillers. The repair is therefore permanent, unless you are unfortunate enough to have another accident at the same place on your car.

you want it. To stop the solder sticking to the paddles, they are dipped into tallow before use. Tallow is mainly used as an industrial lubricant today, but it had a diverse list of uses in the past and will not affect the solder once it is on the panel.

A soft flame is ideal to heat the dollops of solder on the panel. The solder and panel need to come up to temperature slowly, so that once the solidus point is reached, it can be maintained rather than increased. The danger is that if the liquidus point is reached, the carefully placed solder will slip off, so the whole process will need to be gone through again. As the liquidus temperature is reached, the solder will become

---

LEAD LOADING PITFALLS

* Specialist tools needed.
* Extremely clean surfaces required for successful tinning.
* High skill levels are required.
* Too much heat and all the body solder will be on the floor.
* Hand finishing is required to prevent lead dust in the workshop.

Here, a practical demonstration of lead loading is shown. A patch had been welded over a hole in a Mini wing, where a defunct aerial had been. An unsuccessful attempt had been made to hide the repair using plastic filler. This was cleaned out before proceeding to fill with lead, making a more permanent repair.

1. On this Mini wing an old aerial had been removed. The hole had been covered from underneath with a fully welded patch, but had been roughly filled with polyester filler. This was removed back to shiny metal so that the repair could be covered by body solder.

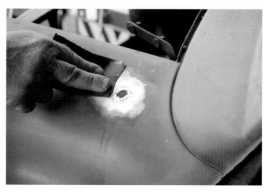

2. Once all the filler had been removed, the area was cleaned further with an emery cloth. An area slightly larger than the repair was prepared, as if left the paint would burn in the heat of the blowlamp.

3. Any rusty pits would require extra work, as it is the thorough cleaning that more or less guarantees a good job. Once this has been done, the repair proper can commence.

4. Using the brush supplied with the tinning solder, an even coat was applied without any gaps. This gave a matt grey appearance to the treated areas.

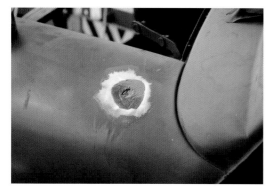

5. An area slightly larger than the repair area needed to be tinned. This would allow the body solder to be blended into the surrounding wing so that the repair would be invisible once painted.

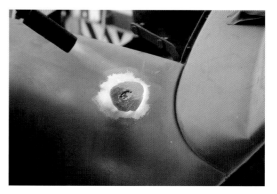

6. The area was heated evenly with the blowlamp flame until the tinning solder melted and took on a shiny silver appearance. This was then wiped off with a clean cloth to remove any excess flux.

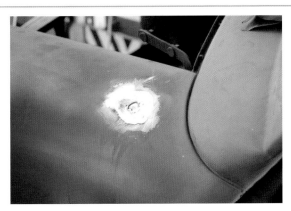

7. The repair area took on a nice shiny appearance, which denoted that the whole area had been tinned and the next stage could be tackled.

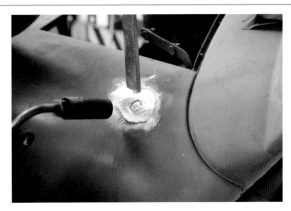

8. The tinned area was heated until the tinning solder had just melted. A twisting motion was used to apply dollops of body solder to the panel. There needed to be enough solder on the panel so that once spread with the paddle it was just proud of the final contour required.

9. If more solder is required, now is the time to add it before spreading the rest. It must not be forgotten that too much heat will cause the solder to slip off the panel.

10. The body solder needed to be kept between the solidus and liquidus points with the blowlamp flame as the solder was manoeuvred into the correct contours. The solder resembled mushy snow as it was pushed around.

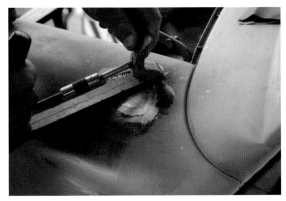

11. After the flux and tallow had been thoroughly washed off with water, the body solder was brought nearer the required shape with the body file. This can be adjusted by the central turn buckle. It will either make the file convex or concave, helping to get the right shape.

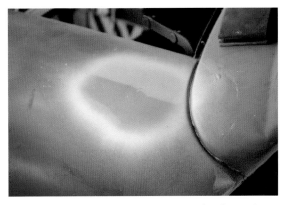

12. Any final shaping can be performed with various grades of wet and dry abrasive paper, working through to the finer grades to achieve a completely blended-in repair before coating with primer. Mechanical sanding should not be used on the body solder as it will inevitably put fine lead particles into the air.

Electrical soldering is quite straightforward once the basics have been learnt. Above all, cleanliness is the key to a well-made joint.

9   Electrical Soldering

Soldering produces a bond at a molecular level with the parent metals. Without going into the theories of electrics and electronics, the only resistances in a circuit should be the ones that are there by design. True, all materials have a resistance, unless we enter the realm of super-cooled conductors, in which the resistance is virtually nil, but this entails vast quantities of cryogenic materials to reduce the temperature of the components down to absolute zero, 0 Kelvin or −273.15°C. In an ideal world, a joint in an electrical circuit should have no resistance, or possibly be the same resistance as what is being joined, in this case usually copper wires to carry electric current. A properly soldered joint conforms to this criterion, as soldering alloys have a good conductance so will not resist the flow of current in the circuit any more than the wires themselves. The joint, being covered from the effects of the atmosphere, that is, oxidation, will be

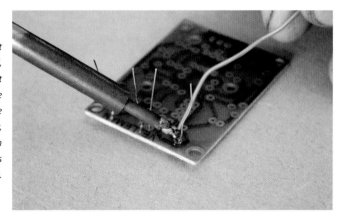

To make a successful joint on a printed circuit board, once the component wire is through the board, approach with the soldering iron and solder, keeping the flat of the iron against the wire, as this will need more heat.

The solder is pushed on to the wire/soldering iron until the solder can be seen flowing on to the printed circuit board.

This is allowed to cool for a second, before moving on to the next joint to be made.

A dry joint takes on the appearance of a plum pudding; the surface tension keeps the solder in a ball.

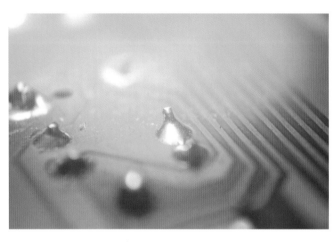

A correctly made joint looks like a volcano, with the solder adhering to the wire and the circuit board.

THE DRY JOINT

When soldering electrical components every effort needs to be made to avoid creating a dry joint, so-named because the solder has not properly wetted both parts to be soldered. Or, such a joint can be created if the components being soldered move before the solder has cooled sufficiently to be completely solidified below the solidus point. This usually leads to increased resistance within the joint, causing a complete failure, or, at the very least, undesired results. These may not manifest straight away, but could happen weeks or months later during service and be difficult to find without checking every joint.

On a circuit board a good soldered joint should take on the appearance of a volcano, wide at the base leading up to a pointed top, with just the cut end of the component wire sticking out. A dry joint will look more like a plum pudding, round at the base as well as at the top, where the solder has not adhered to the copper of the printed circuit board, or the wire of the component. The surface tension of the solder has not broken down, so has kept the solder in a ball, stopping adequate bonding to the other surfaces. This is to be avoided at all costs; either too little heat was applied to the joint, or the

more or less permanent. Any appreciable resistance caused by the soldering would result in possibly unexpected effects in the electronic circuits or excess heat, which would also be detrimental.

In the field of manual connections, the main components used are copper or a copper alloy that lends itself to soldering. In the realms of micro-components, where the contacts on integrated circuits are gold, different materials will be required to make the connections, as normal solders will leach the gold from the contacts, but this chemistry, as interesting as it is, is beyond the remit of this book.

Soldering of electrical and electronic components is a fairly fine balancing act when it comes to applying the heat to make the soldered joint. Obviously heat needs to be applied, but too much and the excess heat

will irreversibly damage the components that we are trying to solder into the circuit; too little and the possibility of the joint failing, now or in the future, due to a dry joint is a real possibility.

After a while, the tip of the soldering iron will need filing to remove any erosion.

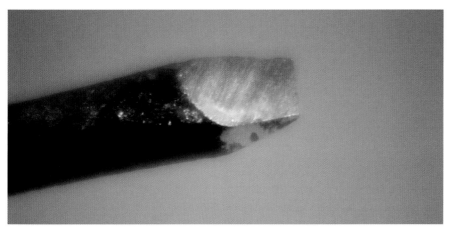

Once the tip is back to clean metal, the soldering iron can be retinned as soon as it is up to temperature.

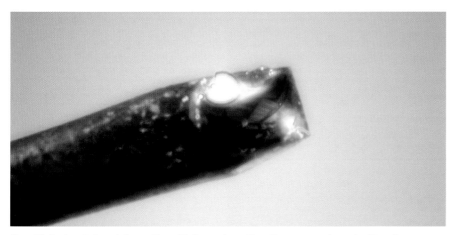

The soldering iron tip is ready for another soldering session, retinned to ensure maximum heat transfer.

copper on the board or the component wire was dirty, or both. Another cause may be an insufficiently tinned iron. For a consistent joint it will pay to get into the habit of tinning the iron just before making contact between the soldering iron and the component. This not only ensures that the iron is adequately tinned, but that it also has fresh flux from the cores within the solder to clean off oxide as the joint is made.

EVEN HEATING

It stands to reason that something large requires more heat input to bring it up to temperature than a small one; so it is with printed circuit boards and components. The copper strips on the board are extremely thin when compared with the component tails. From this statement it can be seen that the component wire will require more heat than the board it is to be attached to during soldering. The technique I use is to place the flat of the tinned soldering iron tip against the component wire and the side resting on the printed circuit board copper. This way, the component wire gets more direct heat and the copper on the board gets enough heat to solder without overheating, which would detach it from the glue holding it to the fibre baseboard. With the solder held in the other hand from the soldering iron, the end of the solder is fed into the tip of the soldering iron and the component wire. The solder will be seen to flow over the component wire and on

to the board with the already mentioned classic volcano outline. The soldering iron is withdrawn at this point and allowed to cool down a bit. After a while, this technique will become second nature, hopefully producing a good joint every time. Described in writing, this all seems a bit long-winded, but in reality it is all completed in a brief moment, with many joints being made in the time it has taken to describe.

Today, the vogue when attaching connectors to the end of wires, particularly on vehicles, is the crimped joint, whereby the connector is squashed on to the wire with specially designed crimping pliers. This type of joint is extremely quick and requires no heat. However, corrosion, especially in a damp environment, can lead to problems not unlike a soldered dry joint, with increases in resistance as oxide layers build up between the wire and the terminal. The remedy to overcome this situation is to solder the joint. This can be done after crimping, giving the best of both worlds. The crimped joint gives mechanical strength and the solder makes a sealed joint, so eliminating the chance of any moisture entering and causing corrosion problems from within the joint. To finish the job off properly, a piece of heat-shrink sleeving should be placed over the joint and extend a couple of centimetres beyond to eliminate any possibility of moisture entering the joint. This is secured by the application of heat to shrink the sleeving tightly against the wire and terminal.

SOLDERING IRONS FOR ELECTRICAL WORK

As seen above, to make a successful joint without damaging components, the correct-sized soldering iron needs to be employed. Too small a soldering iron will take longer to heat the joint components and with this extra time the heat will travel further, possibly damaging sensitive components before enough heat has entered the joint to melt the solder. Too large an iron will possibly

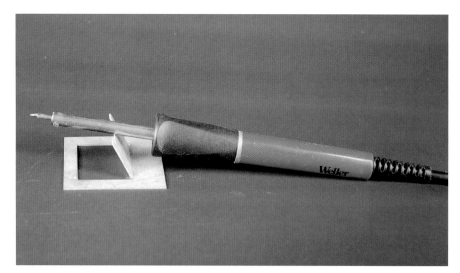

A very small soldering iron is essential when working on tightly packed electronic panels and circuit boards.

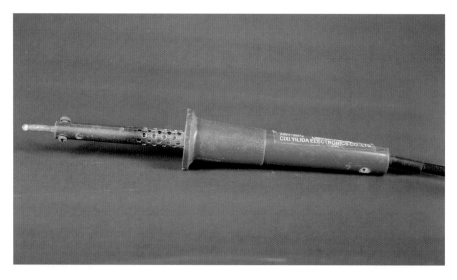

For general electrical soldering, a larger soldering iron will produce a secure joint quickly.

The butane-powered soldering iron is ideal for electrical work, when fitted with a soldering tip.

do the job, but the larger tip on the bigger iron will invariably unsolder joints already made, or the heat radiating from this bigger tip could cause damage to adjacent components.

The normal-sized soldering iron for general electrical work is the 25–30watt iron and for finer work such as on printed circuit boards a smaller iron would be used, such as the 15–20watt iron. This will have a finer soldering tip for easier access among the plethora of connections on a circuit board and heat transfer will be less on the close components.

A butane-powered soldering iron is ideal for use with electrical soldering and of course has the added portability that it does not require mains electricity to power it, so soldering can be performed outside on a broken-down vehicle, for example. With the power being adjustable by varying the volume of gas being burnt, and of course the soldering iron tip being removable, be careful when hot. The gas flame can also be used for shrinking sleeving to give a water-tight finish to the job.

ELECTRICAL SOLDERING FLUX

When soldering electrical components, under no circumstances use a general acid-based flux. As the name implies, the flux uses acid as its active ingredient to clean impurities from the surface being soldered. The problem with these fluxes is that they require washing off after soldering to remove the acid, as slowly but surely they will carry on eating away whatever they were placed on. Water does not mix well with electrical components, so be warned.

MULTICORED SOLDER

Solder that is specifically made for electrical work comes with a flux core, or multi-core, within the solder itself. A naturally occurring family of resins is used – some-times known as Rosin, and sold under

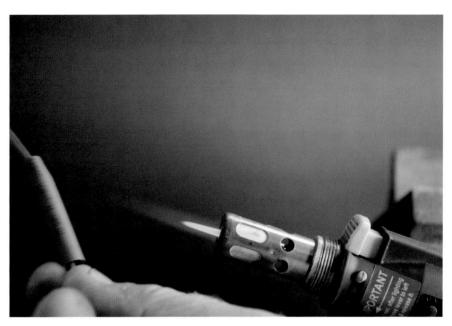

By placing shrink sleeving over the soldered joint that has just been made, it will give insulation and watertightness.

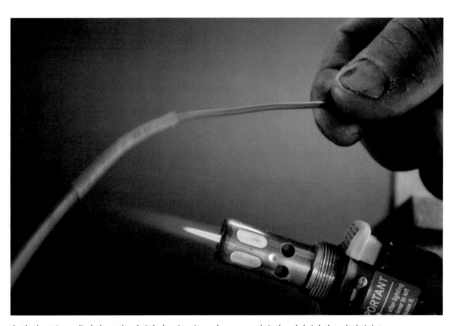

As the heat is applied along the shrink sleeving, it can be seen to shrivel and shrink the whole joint.

trade names such as Ersin – and there are various resin formulae for different solder-ing tasks. At normal temperatures, these resins are practically inert and therefore harmless to the electrical connections made, but at soldering temperatures these fluxes have a mild cleaning effect and, as with all fluxes, eliminate air while the joint is being made. Once cooled and solidified, the remnants of the flux remain on the joint, causing no harm or ill-effects in the short or long term. Various thicknesses of Multicore solder are available, depending on how fine the work is.

For electrical work, the solder usually comes in a handy dispenser, just pulling it out as necessary. This also keeps the solder clean, avoiding contamination.

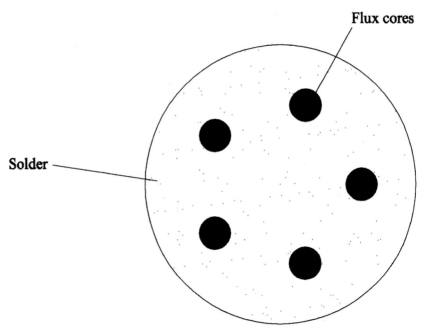

Solder for electrical work is called Multicore, because it has several cores of flux running through it. This allows the right amount of flux to be released as the solder is melted during soldering operations.

CARE OF YOUR SOLDERING IRON FOR ELECTRICAL WORK

As times goes on, your soldering iron will need some care and attention to keep it in good condition and soldering with the minimum of fuss. Although the flux is built into the solder and the correct amount should be administered along with the solder, after a while the soldering iron tip will accumulate a thick brown grungy liquid, which, if left, will solidify as the soldering iron cools down. This is only the flux residue and possibly any oxides that the flux has floated off of the previous job. To remove this messy substance, the easiest thing is have a dampened piece of waste cloth or tissue handy and wipe the iron tip occasionally as you proceed with the work.

Some of the kits sold today for specifically electrical and electronic work come complete with a small metal pot containing a sponge especially for this purpose. This will not only keep the soldering iron tip clean, but will avoid transferring the dross build-up on to the work. This is especially important if doing any fine printed circuit-board work. Although this dross will be fairly inert once cooled down, it will impede

After a while, the tip of the soldering iron gets a build-up of spent flux and oxides. To make work easier the tip should be cleaned from time to time.

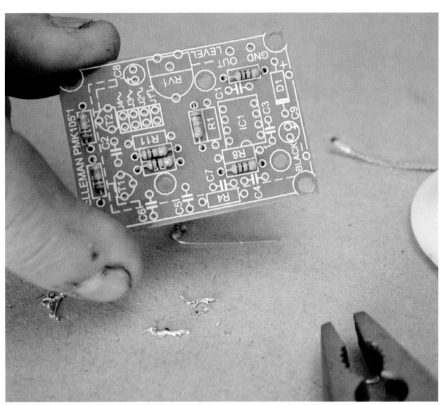

The ideal solution is a purpose-made sponge. While the iron is hot, just wipe on the dampened sponge to remove the build-up.

The printed circuit board, in this case a signal generator. Many components can be crammed into a small space.

your view as you work on such fine close connections.

PRINTED CIRCUIT BOARDS

Where soldering is being carried out on printed circuit boards, or on the Veroboards that are available (these are boards with strips of copper pre-drilled right through the board for component insertion), care has to be taken to avoid damaging the components being placed on the board. Also, too much and prolonged heating will lift the strips from the board base, rendering it useless.

Resistors and capacitors are fairly robust and should stand the heat from soldering them on to the board, it is the components that are made from semi conductor materials, that is, diodes, transistors and the like that are much more heat sensitive. To elimi-

nate this problem special heat sinks that are clipped to the wires of the component are available to shunt the excess heat away while soldering is taking place, before it reaches the sensitive parts of the components; however, a pair of snipe nosed pliers will do an equally good job. The technique is to insert the component through the board and then to grip each wire in turn as it is soldered on the underside of the board (the other side of the soldering operation) so shunting the excess heat into the body of the pliers instead of the component.

Veroboard is very useful for making small circuit boards, but the small thin copper strips are easily pulled from the baseboard. To cut the copper strips, a small twist drill can be used by hand at the appropriate hole.

Some components are more heat-sensitive than others; transistors, diodes and integrated circuits are most prone to damage.

When soldering heat-sensitive components, a heat shunt is used to remove damaging heat before it reaches the component. A pair of sniped nose pliers makes a good heat shunt.

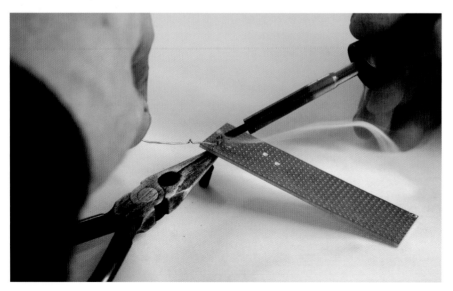

After the component has been placed through the circuit board, the component wire is gripped with the pliers before soldering.

The soldering is carried out normally, with the pliers kept in place for a second or two to drain any residual heat, after the joint has been made.

DESOLDERING

Frequently in repair work it is required to desolder a component or components, so that they can be removed for replacement or perhaps for testing off of the board. Once again, too much heat will be detrimental to the component and the board. Desoldering braid can be used. This is a braid of fine wires with a suitable flux coating, which upon heating will draw the solder into the braid by capillary action away from the joint that you wish to undo. The procedure is simple – all that is required is to place the braid on the joint that is being worked on and heat the joint through the braid. As the solder melts, it will be drawn into the braid. The only downside with this method is that each time it is necessary to move along to a fresh piece of braid.

If a lot of desoldering is to be undertaken, a desoldering pump is available. This is a hand-held device that is cocked before use; once the solder of the joint is liquid under the tip of the soldering iron, the nozzle of the device is quickly placed on the solder and the trigger is released. This allows the spring-loaded plunger inside of the tube of the device to draw up the molten solder, leaving the joint clear. The nozzle is made from a heat-proof material to which the solder will not adhere. The end is removable to entail the emptying of the spent solder every so often, otherwise the device would soon clog with solidified solder, rendering it useless.

Desolder braiding is manufactured in a number of sizes. To use, hold on the joint that you want to desolder and place the hot soldering iron on to the braiding. The solder on the joint will be drawn into the fluxed braiding.

The desoldering pump is ideal for removing solder from joints. To use, heat the joint with the soldering iron and then release the cocked pump and it will suck the molten solder from the joint. After a while the pump will start to fill with solidified solder; the pump comes apart to empty.

MODERN WONDERS

As mentioned above, during the connection of a wire end terminal with solder for automotive use, the connection was finally covered with a length of shrink sleeving. This is really clever stuff. It comes in all manner of sizes and colours and is insulating. Although the soldered joint itself is relatively safe from corrosion from the elements, where the wire extends from its sleeve, the multistranded wire will be prone to the ingress of moisture by capillary action up into the sleeve. Out of sight, the moisture and any detrimental atmospheric compounds will slowly but surely corrode the wires, giving problems in the future. The act of sealing this wire/sleeve interface with shrink sleeving will eliminate the problem and, as it is insulating in itself, will require no additional work to ensure adequate insulation.

WIRE ENDS

Most, if not all, mains electrical items come with a plug already fitted, suitable for the local voltage. If, however, the plug has been removed for whatever reason, or requires replacement due to damage, it is a good idea before placing the wires under the correct terminals in the new plug to solder the twisted wires. This has the effect of holding the multistrand wires together as one, giving a much better connection as the terminal screw is tightened. The connection will then be much longer lasting than just tightening the screw on the wire strands, where they can splay apart and weaken the connection. Also, if all the strands are not clamped, the full current-carrying capacity will not be utilized, possibly leading to over-heating in the appliance plug. This in itself can cause the working loose of the terminal screw, which is not only annoying, but will also exacerbate the situation.

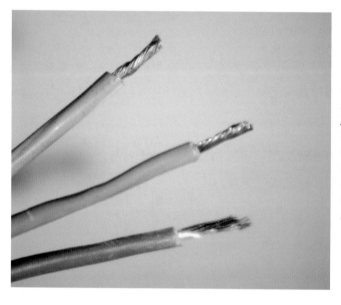

When rewiring a mains plug it makes a better job if the wires are tinned before fitting in the plug; live wire (brown) as bared, neutral wire (blue) twisted and the earth wire (yellow/green) tinned ready to be fitted.

Assorted crimped connectors; red for wires up to 1.5mm², blue 2.5mm² and yellow for 4mm².

Soldering a crimped connector after removing plastic insulation. Once soldered, the joint can be fitted with shrink sleeving to protect the joint and reinstate the insulation.

EARTHING ISSUES

Although this book shows only the rudiments of electrical soldering, it would be incomplete if the importance of earthing during this type of work was not mentioned. In the past, anything plugged into the mains electricity had an earth wire. This was there so that if anything shorted to the metal case of the appliance, the earth wire would cause enough current to blow the fuse, rendering it safe, as it is at the same potential as the neutral wire. Thus a severe electrical shock, or even death, could be avoided. However, with modern advances in insulation and insulating materials, this earth has been dispensed with and a lot of small appliances will only have two wires to them, a live and a neutral. The appliance will have an internationally recognized, double-insulated logo on it somewhere, consisting of two squares one inside of the other.

This has proved quite safe to avoid any shocks from the mains voltage of 240volts, but when working with sensitive components, especially integrated circuit chips, there is another problem, static electricity. The tip of the soldering iron is usually earthed, in which case the static issue is not too much of a problem here, but it will be acerbated by the modern trend of nylon carpets, or rubber-soled shoes, by which static electricity is easily generated. In addition, the now dry air of the centrally heated house does not help to dissipate any static charge. The human body will act as a store for any static build-up if insulated by wearing rubber-soled shoes and this will need to be drained, otherwise damage could occur to sensitive components. The operator may not even notice the discharge, although sometimes in a room with a nylon carpet when touching an earthed object such as a fire that has an earth connection, some spectacular sparks can be produced, combined with quite a jolt from the several thousand volts that have built up. This will do you no harm, other than make you jump, if you feel it at all. Although the voltage could be high, possibly 2,000–3,000 volts, the current will be extremely small, and it is the current, not the voltage, that will kill you. Electronic components, however, will not withstand these sorts of voltage spikes.

The remedy is to make sure that you are earthed and continue to be earthed as you work and not build up any more static, caused by rubbing your shoes on the carpet.

One way to discharge yourself of static is to touch something regularly that is earthed where you are working, such as a radiator or pipework to it. A better way is to purchase one of the specially made earthing wristbands. These come with a flexible wire, or carbon fibre, that clips to a good earth point. This has the effect of draining any static from your hands right where you are working, before it can build up, thus protecting your components – and of course your temper and inevitably your bank balance when components are fried!

PITFALLS WHEN SOLDERING ELECTRICAL CONNECTIONS

* Never use any acidic flux intended for plumbing work on electrical connections.
* Avoid producing dry joints.
* Do not move wires before the solder has solidified.
* Use the correct size of soldering iron for the job.
* Never heat the joint longer than necessary.

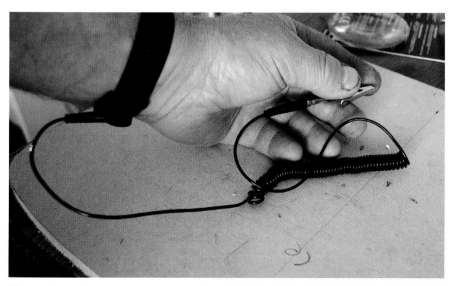

When working with sensitive components, an earth wristband will take away any static electricity. The crocodile clip should be attached to a good earth point, such as a central heating pipe.

It was discovered that the wires leading to the anti-lock brake sensor on this car were faulty. Closer examination showed that a previous repair had failed, as the broken wires were just twisted together. The bare ends of the wires were scraped clean and tinned before soldering the correct wires together and covering with heat-shrink sleeving, giving a more permanent repair.

Sometimes a large enough soldering iron is not available for the job in hand. In pictures 7 to 12, a demonstration is shown soldering a large battery cable terminal using just a gas blow torch.

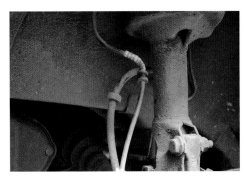

1. A previous repair on the car's anti-lock braking system had broken down, triggering the warning light on the dashboard. This is a common problem where the wires are in constant movement, causing them to break through fatigue.

2. After removing all the old insulation from the two wires, these were cleaned back to shiny copper so that the solder would tin them correctly. The two wires from the sensor were cleaned as well.

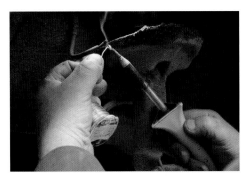

3. Using a hot soldering iron and electrical solder, each wire was heated and solder applied to the wire itself. If clean, the solder will flow into the wire, correctly tinning them.

4. Before the wires were soldered together, lengths of heat-shrink sleeving were slipped on to each one.

5. Heat was applied, in this case a lighter, until the sleeving had shrunk tightly to the wires.

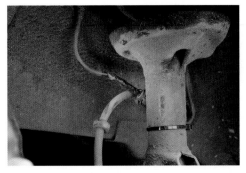

6. Ideally, a larger piece of heat-shrink sleeving would have been slid over the outer wire cover to complete the job, but I discovered that all the larger pieces had been used, so a tight wrapping was achieved with two layers of insulation tape. After clipping the plug back on the sensor and securing the wire with a cable tie, a test drive proved the job, as the warning light went out.

7. Fitting a terminal to a starter motor cable is not difficult, just more heavy duty than normal electrical soldering.

8. After the insulation had been stripped back, the strands of wire needed to be tinned. A large soldering iron should be used, although it is possible to do it with just a blowtorch.

9. Once tinned, the new terminal was fitted to the tinned wire, with the two tabs folded in nice and snug. The terminal was zinc-plated, so soldered easily.

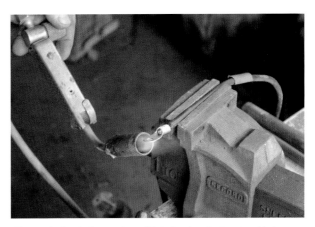

10. The terminal and wire were heated, bringing them both up to soldering temperature. Too much heat on the insulation should be avoided.

11. The electrical solder (flux cored) was now fed in. It can be seen filling in any gaps around the terminal; more heat should be added if necessary until the terminal end is filled.

12. The soldered joint was now covered with heat-shrink sleeving. This not only looks neat, it will also prevent the ingress of moisture into the wire itself. Everything was now ready to be fitted back on the engine from which it had come.

Useful Addresses

Below is a list of some useful addresses of where equipment, alloy data and safety gear can be purchased and further information sought. This list is only a starting point, and I have no connection with any of the companies mentioned:

Air Products
Suppliers of industrial gases.
Air Products PLC
2 Millennium Gate
Westmere Drive
Crewe
Cheshire
CW1 6AP
Phone 0800 389 0202

BOC
Nationwide network of branches supplying industrial gases and welding equipment; an account is required to purchase gases but will sell ancillary equipment over the counter to non-account holders.
BOC Customer Services Centre
PO Box 6
Priestley Road
Worsley
Manchester
M28 2UT
Phone 0800 111 333, www.boconline.co.uk

Bullfinch Gas Equipment
Manufacturer of gas torches and associated equipment.
Diadem Works
Kings Road
Tyseley Birmingham B11 2AJ
Phone 0121 765 2000, Fax 0121 707 0995, www.bullfinch-gas.co.uk

Calor Gas Ltd
Nationwide agents supplying LPG gas.
Calor Gas Ltd
Athena House
Athena Drive
Tachbrook Park
Warwick
CV34 6RL
Phone 01926 330 088, Technical Help Desk 0845 602 1143, www.calor.co.uk

CUP Alloys of Chesterfield
Suppliers of solder, silver solder and brazing alloys, along with a host of ancillary equipment and specialized knowledge.
Phone 01909547248, www.cupalloys.co.uk

Frost Auto Restoration Techniques Limited
This company supplies restoration tools mainly for the classic car enthusiast, but has some useful hard to find tools for the budding welder.
Frost Auto Restoration Techniques Limited
Crawford Street
Rochdale
Lancashire
OL16 5NU
Phone 01706 860 338, www.frost.co.uk

Hobbyweld Gas
This company supplies a nationwide chain of retailers that supply shielding gases and oxygen on the payment of deposit for the first cylinder, after which the only payment is for the gas purchased, no matter how long you have the cylinder. Once finished, on return of the cylinder the original deposit payer can reclaim the deposit.
Hobbyweld Gas
Dixons of Westerhope
Westfield
New Biggin Lane
Westerhope
Newcastle-upon-Tyne
NE5 1LX
Phone 0800 433 4331, www.hobbyweld.co.uk

Johnson Matthey
Manufacturer of solders and brazing alloys.
Johnson Matthey PLC
Metal Joining
York Way
Royston
Hertfordshire
SG8 5HJ
Phone 01763 253200, Fax 01763 253168, www.jm-metaljoining.com

Machine Mart
Has branches all around the country where its products can be viewed and purchased or ordered online from its colour catalogue.
211 Lower Parliament Street
Nottingham
NG1 1GN
Phone 0844 880 1250, www.machinemart.co.uk

Northern Tool +Equipment

Suppliers of hand tools and machinery.

Northern Tool + Equipment

Anson Road

Portsmouth

PO4 8TB

Phone 0845 605 2266,

www.NorthernToolUK.com

Sievert UK Ltd

Manufacturers of LPG blowtorches.

Sievert UK Ltd

Bridge Street

Holloway Bank

Wednesbury

West Midlands

WS10 0AW

Phone 01215061810, Fax 08454588590 or

www.sievert.co.uk

Index

Related Titles from Crowood

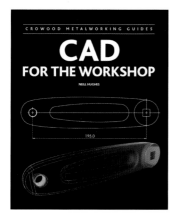

NEILL HUGHES
ISBN 978 1 84797 566 9
112pp, 210 illustrations

DR MARCUS BOWMAN
ISBN 978 1 84797 512 6
144pp, 280 illustrations

CHRIS TURNER
ISBN 978 1 84797 776 2
192pp, 230 illustrations

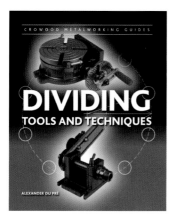

ALEXANDER DU PRÉ
ISBN 978 1 84797 838 7
144pp, 240 illustrations

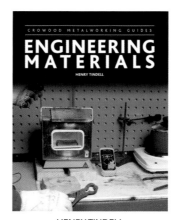

HENRY TINDELL
ISBN 978 1 84797 679 6
192pp, 230 illustrations

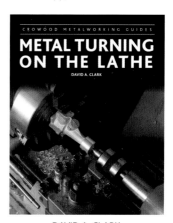

DAVID A. CLARK
ISBN 978 1 84797 523 2
112pp, 240 illustrations

DAVID A. CLARK
ISBN 978 1 84797 774 8
160pp, 210 illustrations

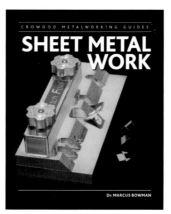

DR MARCUS BOWMAN
ISBN 978 1 84797 778 6
160pp, 450 illustrations

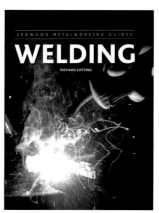

RICHARD LOFTING
ISBN 978 1 84797 432 7
160pp, 280 illustrations

www.crowood.com